Convergence of Blockchain and Explainable Artificial Intelligence

BlockXAI

RIVER PUBLISHERS SERIES IN DIGITAL SECURITY AND FORENSICS

The "River Publishers Series in Security and Digital Forensics" is a series of comprehensive academic and professional books which focus on the theory and applications of Cyber Security, including Data Security, Mobile and Network Security, Cryptography and Digital Forensics. Topics in Prevention and Threat Management are also included in the scope of the book series, as are general business Standards in this domain.

Books published in the series include research monographs, edited volumes, handbooks and textbooks. The books provide professionals, researchers, educators, and advanced students in the field with an invaluable insight into the latest research and developments

- Blockchain for secure transactions
- Cryptography
- Cyber Security
- Data and App Security
- Digital Forensics
- Hardware Security
- IoT Security
- Mobile Security
- Network Security
- Privacy
- Software Security
- Standardization
- Threat Management

For a list of other books in this series, visit www.riverpublishers.com

Convergence of Blockchain and Explainable Artificial Intelligence

BlockXAI

Editors

Akansha Singh

Bennett University, Greater Noida, India

Krishna Kant Singh

Delhi Technical Campus, Greater Noida, India

River Publishers

Routledge
Taylor & Francis Group
NEW YORK AND LONDON

Published 2024 by River Publishers
River Publishers
Alsbjergvej 10, 9260 Gistrup, Denmark
www.riverpublishers.com

Distributed exclusively by Routledge
605 Third Avenue, New York, NY 10017, USA
4 Park Square, Milton Park, Abingdon, Oxon OX14 4RN

Convergence of Blockchain and Explainable Artificial Intelligence / by Akansha Singh, Krishna Kant Singh.

Routledge is an imprint of the Taylor & Francis Group, an informa business

ISBN 978-87-7004-175-1 (hardback)
ISBN 978-87-7004-647-3 (paperback)
ISBN 978-87-7004-639-8 (online)
ISBN 978-87-7004-630-5 (master ebook)

Contents

Preface

Explainable AI (XAI) is an upcoming research field in the domain of machine learning. The existing AI systems make decisions in a black-box manner. XAI explains how an AI system arrived at a particular decision. It inspects the steps and models that are responsible for deducing a particular decision. It is an upcoming trend that aims at providing explanations to the AI decisions. Blockchain is emerging as an effective technique for XAI. It enables the accessibility to digital ledgers amongst the various AI agents. The AI agents collaborate using consensus and decisions are saved on blocks. These blocks can be traced back but cannot be changed. Thus, the combination of AI with blockchain provides transparency and visibility to all AI decisions. BlockXAI is also being widely used for improving the data security and intelligence. The decisions made are consensus based and decentralized leading to highly efficient AI systems. This book aims to provide a detailed description on the topics related to XAI and blockchain. These two technologies can benefit each other. The research outcomes will benefit the society in multiple ways. It will also cover topics that present the convergence of blockchain with explainable AI. This book will provide the researchers, academicians, and industry experts a complete guide to BlockXAI.

List of Figures

List of Tables

List of Contributors

Abouhawwash, Mohamed, *Department of Mathematics, Faculty of Science, Mansoura University, Mansoura, Egypt*

Balaguru, S., *JKK Nataraja Engineering College, India*

Bharti, Anuj Kumar, *School of Computer Science Engineering and Technology, Bennett University, India*

Cengiz, Korhan, *Trakya University, Turkey*

Husain, Mohammad, *Faculty of Computer and Information Systems Islamic University of Madinah, Saudi Arabia*

Izonin, Ivan, *Department of Artificial Intelligence, Lviv Polytechnic National University, Ukraine*

Kalimuthu, Sivanantham, *Crapersoft, India*

Koti, Manjula Sanjay, *Department of MCA Dayananda Sagar Academy of Technology & Management, Bangalore, India*

Kumar, R. Mohan, *Sri Ramakrishna Engineering College, India*

Kumar, S. Satheesh, *Department of Computer Science, School of Applied Sciences, REVA University, India*

Kumar, V. Vinoth, *School of Computer Science Engineering and Information Systems, Vellore Institute of technology, Vellore, India.*

Mahalakshmi, J., *Crapersoft, India*

Muthukumaran, V., *Department of Mathematics, College of Engineering and Technology, SRM Institute of Science and Technology, Kattankulathur - 603203, Tamilnadu, India*

Praveen, P. Blessington, *Crapersoft, India*

Raghuvanshi, Shikhar, *School of Computer Science and Engineering, Vellore Institute of Technology, India*

Saravanan, Vijayalakshmi, *University of South Dakota, Department of Computer Science, USA*

Singh, Akansha, *School of Computer Science Engineering and Technology, Bennett University, India*

Singh, Krishna Kant, *Delhi Technical Campus, India*

Srivastava, Himanshu, *School of Computer Science Engineering and Technology, Bennett University, India*

Swapna, B., *Department of Electronics and Communication Engineering, Dr.M.G.R. Educational & Research Institute, India*

Tiwari, Raj Gaurang, *Chitkara University Institute of Engineering and Technology, Chitkara University, India*

Tyagi, Amit Kumar, *National Institute of Fashion Technology, New Delhi, 110016, India*

Viswanathan, Arun, *School of Computer Science and Engineering, Vellore Institute of Technology, India*

List of Abbreviations

ABE	Attribute-based encryption
ACAS	Access control aware search technique
ADE	Adverse drug event
AES	Advanced encryption standard
AI	Artificial intelligence
ANN	Artificial neural network
AR	Augmented reality
AVI	Audio video interleave
BFP	Binary floating point
BlockXAI	Blockchain and XAI
BPMH	Best possible medication history
CAD	Computer-aided design
CDSS	Clinical decision support system
CoAP	Constrained application protocol
CSHSC	Center for the study of health system change
CVD	Cardiovascular disease
DeFi	Decentralized finance
DLL	Dynamic link library
DoS	Denial-of-service
DTLS	Datagram transport layer security
DVT	Deep vein thrombosis
EDA	Electronic design automation
EHR	Electronic health record
EMR	Electronic medical record
HCI	Human-computer interaction
HER	Electronic health record
HIoT	Healthcare IoT
HIS	Health information system
HRD	High required dose
HSC	Health services commission
INR	Internationalized normalized ratio

IoT	Internet of things
IoV	Internet of vehicles
IPFS	InterPlanetary file system
IWPC	International warfarin pharmacogenetics consortium
LIME	Local interpretable model-agnostic explanations
LRD	Low required dose
LRP	Local regression planning
MDP	Markov decision processes
ML	Machine learning
MQTT	Message queue telemetry transport
NEC	Narrow elliptical curve
NECPSO	Narrow elliptical curve with particle swarm optimization algorithm
NLP	Natural language processing
NN	Neural network
P2P	Peer-to-peer
PAI	Physical activity index
PCA	Principal component analysis
PDF	Portable document format
PDP	Partial dependence plot
PJE	Pointing and reasoning explanation
PoS	Proof of stake
PoW	Proof of work
PT	Prothrombin time
SaaS	Software as a service
SC	Smart contract
SDN	Software-defined networking
SHAP	Shapley additive explanations
SSL	Secure sockets layer
TCP	Transmission control protocol
TSP	Telecommunication service provider
UDP	User datagram protocol
UEM	Unified endpoint management
UIMS	User interface management system
UX	User experience
VR	Virtual reality
XAI	Explainable artificial intelligence

1

BlockXAI: Review of Blockchain for Explainable Artificial Intelligence

Himanshu Srivastava[1], Akansha Singh[1], Anuj Kumar Bharti[1], and Korhan Cengiz[2]

[1]School of Computer Science Engineering and Technology, Bennett University, India
[2]Trakya University, Turkey
Email: akanshasing@gmail.com

Abstract

In recent times, artificial intelligence (AI) has found applications across various domains, notably in critical sectors like healthcare and the stock market. The increasing reliance on AI in these areas has underscored the need for explainable artificial intelligence (XAI), where users expect not just results but also explanations for AI-driven decisions. This demand has spurred the development of XAI, which provides descriptive rationales for specific outcomes.

A significant challenge in the implementation of XAI is its integration into decentralized environments, where blockchain technology emerges as a promising solution. Known for its security and non-repudiation features, blockchain offers a robust and secure method for implementing XAI. This chapter outlines the application of blockchain in developing an XAI framework, reviewing existing literature on both technologies and exploring the concept of smart contracts for AI predictors using blockchain. The synergy of blockchain and AI, termed BlockXAI, promises a secure and efficient approach to developing reliable XAI systems.

This review emphasizes the current importance of XAI, its implementation via blockchain, potential application areas of BlockXAI, the benefits of using blockchain for XAI, and future directions for the convergence of blockchain and XAI.

Keywords: Blockchain, Cryptographic Hash, Consensus Protocol, Smart Contract, Explainable Artificial Intelligence, Machine Learning.

1.1 Introduction

The exponential growth in the interaction of big universal objects, big data, and graphical processing units has significantly empowered AI in numerous fields. AI encompasses tasks performed by programs or machines that require intelligence, and its subfields include machine learning, natural language processing, computer vision, and deep learning [1]. AI's core characteristics are algorithms, computing power, and data. With the advent of big data, diverse data formats from various fields pose challenges, often leading to centralized systems for training and analysis. However, centralized systems are vulnerable to single-point failures and adversarial attacks [2].

Current AI models often operate as black boxes, obscuring the intermediate steps of their training processes. As AI algorithms grow more complex, it becomes challenging for users to understand how these models arrive at specific decisions [3]. This lack of transparency has led researchers to focus on making AI more explainable and trustworthy [4, 5]. Blockchain technology, gaining momentum for its security applications, is increasingly being used to develop frameworks for explainable AI in decentralized environments [6].

This chapter presents a systematic review of blockchain for XAI frameworks, offering an overview of blockchain, explainable AI, and their integration for improved outcomes in decentralized networks. It also discusses various applications and benefits of BlockXAI, structured into five sections.

1.2 Blockchain for XAI

1.2.1 Blockchain overview

Blockchain has emerged as a globally recognized term, symbolizing both social commitment and a new era of trusted technology. Initially proposed as a record-keeping system for Bitcoin, a cryptocurrency, it has since expanded its applications to various fields that traditionally rely on intermediaries to establish trust. Centralized systems, prevalent in transactions such as money transfers, individual identity verification, land record management, polling, and intellectual property rights, often depend on third parties. Blockchain technology, however, offers a paradigm shift by reducing reliance on these intermediaries. It enables individuals to establish mutual trust while preserving their privacy [7]. Consequently, blockchain is increasingly being adopted as a more reliable and trusted system for record-keeping, effectively eliminating the need for middlemen in various sectors. A blockchain is an open-source, immutable, tamper-proof, publicly available, and decentralized digital ledger. It consists of a series of blocks linked together to form this ledger, which is collectively maintained and agreed upon by all network participants [7]. Each block in the blockchain stores records or transactions, along with a hash value and timestamp. The blocks are interconnected, with each block linked to its preceding block via a hash value. The immutability of blockchain, a critical characteristic, is achieved through the use of cryptographic hash functions. As depicted in Figure 1.1, the data within a block is processed using a cryptographic hash function (fH), generating a unique hash value for each new block. Should any alteration be attempted on a block, the hash value changes, breaking the chain's continuity and alerting all participants.

Blockchain technology also employs the concept of smart contracts (SC). These are programmable codes capable of self-execution and self-authentication, and they enforce the terms of a contract [8]. Smart contracts facilitate interactions among parties within a network, executing transactions between untrusted parties without the need for a middleman. These transactions are irreversible and traceable, with all participants required to agree on the SC terms in advance. Smart contracts are deployed on the blockchain using programming languages, and they are automatically triggered when predefined conditions are met, eliminating the need for manual paperwork. Ethereum [9] and Hyperledger Fabric [10], which supports Java and Go languages, are popular platforms for creating smart contracts.

Trust and coordination among nodes are essential in blockchain networks. Consensus mechanisms address these issues in decentralized networks [11]. There are several consensus protocols, with the main types being Proof of Work (PoW) and Proof of Stake (PoS). The PoW protocol, proposed by Satoshi Nakamoto [7], involves miners competing to solve complex mathematical problems to create new blocks, with successful miners rewarded in cryptocurrency. However, PoW is energy-intensive and can cause delays in large networks, making it less suitable for real-time AI applications [12]. PoS, a newer protocol, addresses these energy consumption issues [13] and allows stakeholders to create new blocks and increase their stake.

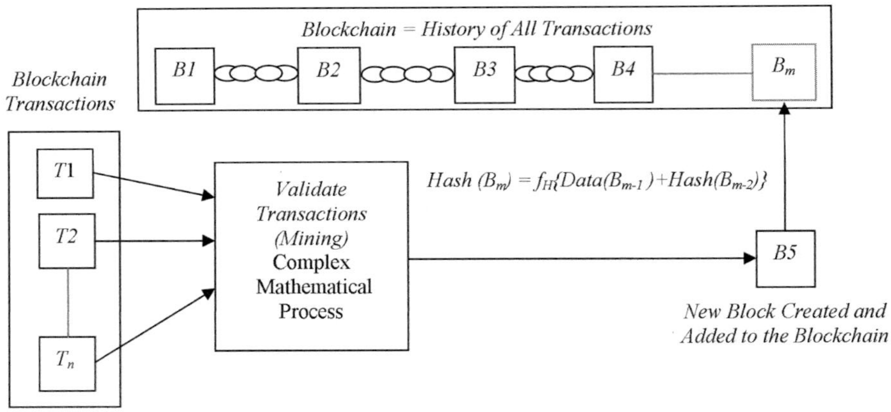

Figure 1.1 Block creation and addition in a blockchain.

Blockchain can be categorized into public, private, and consortium blockchains [14]. Public blockchains, or permissionless chains, allow unrestricted participation and transactions. Cryptocurrencies like Bitcoin, Ethereum, and others typically use permissionless blockchains. Private, or permissioned, blockchains are controlled by specific organizations, restricting node participation. Consortium blockchains, a hybrid of public and private, pre-select nodes for participation in the consensus process, with Hyperledger Fabric being an example of such a blockchain.

Traditional blockchains have limitations in data storage capacity and cost. For instance, the Bitcoin blockchain can only store 1 megabyte of data per block [7]. To overcome this, decentralized data storage systems like the InterPlanetary File System (IPFS) [15], Storj [16], Filecoin [17], BigChainDB [18], and Swarm [19] are used. These systems, like IPFS, are decentralized peer-to-peer networks sharing a single file system, where records are addressable via hash addresses. Their decentralized nature ensures reliability and accessibility, even if a single device in the network becomes disconnected.

1.3 Limitations of Artificial Intelligence

Artificial intelligence (AI) systems currently face several limitations, including sensitivity to biased data, vulnerability to cyber-attacks, and issues with explainability. While AI systems perform adequately with subsets of data, they often struggle with larger datasets due to the inherent biases in the data [20]. Additionally, AI systems are susceptible to cyber-attacks, where malicious actors may manipulate data, leading to incorrect decisions by the AI model.

A significant challenge in AI implementation is its "black-box" nature, which lacks transparency and human interpretability [21]. In a black-box model, the internal workings of the algorithm remain opaque, even to the engineers or scientists who develop them. This opacity makes it difficult for users to understand how the model is trained and how it arrives at specific outputs. As a result, there is a growing need for trustworthy explanations in AI decision-making, especially as biases and attacks on AI models become more prevalent.

AI has gained widespread acceptance and is implemented in numerous fields. However, the lack of transparent explanations for decisions made by AI models poses a significant challenge. In some applications, such as gaming, object recognition, and chatbots, the outcomes of AI models might be accepted without question. Yet, this lack of transparency is unacceptable in sensitive applications involving finance, healthcare, security, and societal issues [6], where the implications of AI decisions can be far-reaching and impactful.

1.4 Integration of Blockchain and XAI

The integration of machine learning algorithms into society necessitates interpretability to foster societal acceptance. Generally, providing explanations is a social process that enhances interactions. Miller [22] defines interpretability as "the degree to which a human can understand the cause of a decision." A highly interpretable model is one that allows individuals to easily comprehend its conclusions or predictions. An AI model is considered well-explainable when its outcomes are more easily understood by humans compared to those from other models.

Elon Musk, a leading figure in technology, has stated that AI technology must ultimately result in either a net positive or negative impact [23]. This outcome depends on how the technology is developed and utilized by humans. Explainable artificial intelligence (XAI) aims to incorporate ethics into AI, steering it toward a positive direction. XAI challenges the notion of AI as merely a black-box model. It represents a self-informative AI model that elucidates the reasoning behind its predictions and decisions [24]. While traditional black-box AI algorithms work on data to produce outcomes, XAI predictors differ in their processing methodology by providing reasoning alongside decision outcomes. The clarity offered by an XAI system enhances user experience through a better understanding of the model's output [25].

Artificial intelligence faces challenges related to trustworthiness, transparency, interpretability, and security. In contrast, blockchain technology offers decentralization, explainability, coordination, transparency, trustworthiness, non-repudiation, and immutability [6]. Integrating these two technologies is essential for creating a new digital world where AI can be confidently utilized. AI models built on blockchain can achieve enhanced security, privacy, and efficiency.

The concept of decentralized AI, which combines blockchain and artificial intelligence, is gaining traction [26]. Decentralized AI makes decisions using trustworthy, secure, and cryptographically signed data, eliminating the need for intermediaries [27]. This approach addresses the issues of black-box AI models by transitioning from centralized to distributed AI agents, providing explainable decisions that are permanently and securely recorded on the blockchain or decentralized storage.

Blockchain technology augments decentralized AI systems by employing a publicly accessible and immutable digital ledger, distributed among AI agents across the network [26]. This setup enables AI agents to reach consensus and record conclusions on the blockchain, ensuring that these decisions are trackable and unalterable by all participants [28]. Smart contracts (SCs) based on blockchain programming models ensure the self-execution of AI agents according to pre-established clauses and agreements [29]. The amalgamation of blockchain with AI systems enhances data and intelligence security through a consensus-based approach [30].

In light of recent discussions on XAI and its integration with blockchain technology, this chapter presents a blockchain-based architecture for explainable artificial intelligence. As illustrated in Figure 1.2, this architecture employs a voting system among multiple XAI agents or predictors operating in a distributed environment. These agents provide outcomes with explanations, gradually building a reliability rating for each predictor. The reliability of a result is significantly enhanced by its explanation, which is audited in a tamper-proof, immutable, append-only, and decentralized digital ledger, ensuring trustworthy tracking. Figure 1.2 depicts a general framework for a blockchain-based XAI system.

The components of the blockchain as depicted in Figure 1.2 are as follows:

1. **XAI predictors:** These are classifiers that employ various classification techniques. The model consists of "*N*" number of XAI predictors or

agents. These agents perform computations on data and provide outcomes along with explanations. The results and their corresponding explanations are audited and recorded in the blockchain, which are then used in subsequent decision-making processes.

2. **Blockchain platform:** This includes both the blockchain itself and smart contracts. The blockchain is used for storing data and decisions. Due to the impracticality of storing vast amounts of data on the blockchain, this model utilizes decentralized data storage systems, such as IPFS, which are integrated with blockchain technology for data storage. Smart contracts facilitate communication among network participants. These contracts, written in programming languages, are executed when predefined conditions are met.
3. **Registration services:** Before participating in the framework, XAI predictors must undergo a registration process. Post-registration, users can select any predictor based on its performance rating.
4. **Rating services:** XAI predictors receive incentives or rewards for reporting accurate outcomes. Conversely, penalties are imposed for inaccurate reporting, which can negatively impact their ratings. This model incorporates a rating system that serves as a reward/penalty mechanism. Users can improve or degrade the rating of a model based on its performance. The rating of an XAI predictor influences user choice for specific tasks.

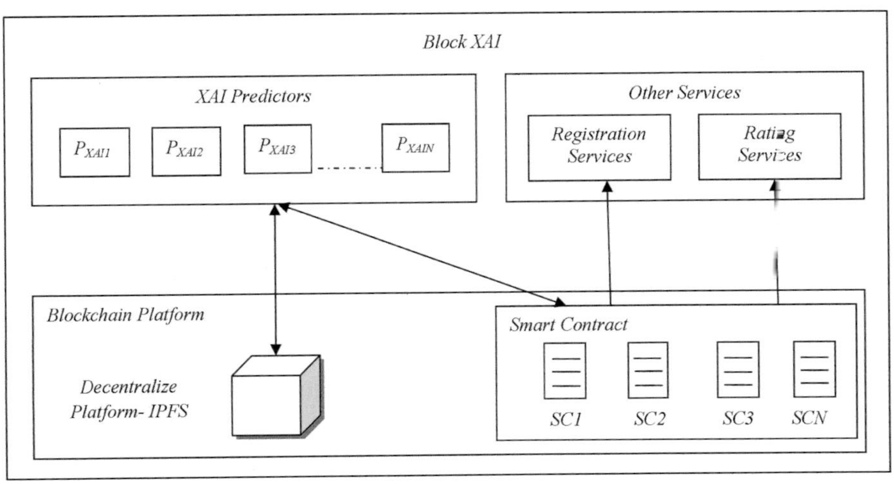

Figure 1.2 BlockXAI architecture.

XAI predictors are capable of producing two types of outcomes: deterministic and non-deterministic. Deterministic outcomes are definitive, yielding results such as yes/no, positive/negative, or true/false. Non-deterministic outcomes, on the other hand, are probabilistic and less precise [6]. To achieve clear, descriptive, and exact outcomes, BlockXAI employs a consensus mechanism among different predictors, as shown in Figure 1.2.

1.5 Application Area of BlockXAI

AI technology, when combined with blockchain characteristics, finds applications in numerous domains. A notable example is automated credit risk assessment, a critical AI application. Traditional AI models, however, face challenges in privacy, security, and interpretability of data [31]. The reliability of these models hinges on human interpretability or the ability to explain outcomes. Blockchain is employed in such scenarios to securely store and audit explanations of AI-generated results, particularly in credit scoring and ensuring data privacy and integrity.

Another area closely linked to public health safety is vaccine and medicine safety. Issues like expiration and record-keeping fraud necessitate a robust vaccine regulatory system to enhance the medical supply chain process. BlockXAI can play a pivotal role in this context [31], where blockchain builds user trust and AI provides data analysis methods.

The increasing use of electric cars, crucial for environmental preservation, highlights the need for clear information about charging services provided by governments and private companies. A blockchain-based charging platform can offer transparent and convenient services by facilitating collaboration among different charging station companies [32].

Food supply management, involving multiple stakeholders like farmers, transporters, processors, and distributors, is prone to fraudulent activities due to uneven information flow. BlockXAI can ensure food safety in supply chain management by providing a transparent and secure system [33].

In the societal domain, polling management is vital for ensuring quality and fairness. The lack of transparency and secure auditability in current voting systems can be addressed with a blockchain-based electronic voting system [34], ensuring tamper-proof and publicly verifiable digital voting.

The rapid development of IoT and the consequent massive data generation have led to increased energy expenditure. A blockchain-based model has been proposed to address this issue by reducing energy consumption [35]. In smart transportation, blockchain can enhance trust, security, and reliability, crucial for developing the Internet of Vehicles (IoV) [36].

The healthcare industry, especially post-COVID, requires secure handling of personal information. Sharing patient data among medical institutions for accurate treatment is a challenge that BlockXAI can address. Blockchain provides a trustworthy mechanism for data security, while AI offers insights through explainable data analysis [37].

To date, various applications have benefited from the integration of blockchain and AI, known as BlockXAI. Its use is crucial for creating a smart and trustworthy society. The potential for BlockXAI in numerous other applications is vast and likely to be explored in the near future.

1.6 Benefits of BlockXAI

Blockchain technology offers several key features that enhance the capabilities of explainable artificial intelligence (XAI):

1. **Enhanced data security:** Blockchain encodes transaction records, making it challenging for hackers to compromise the system. Each block is linked to both its preceding and subsequent blocks in the digital ledger. To alter a single record, a hacker would need to modify the entire blockchain [38]. This robust security framework allows AI models to work with secured data, leading to more trustworthy outcomes.
2. **Trustworthy machines:** The value of an AI model's decision is contingent on its ability to explain that decision in a way that users can understand. Blockchain excels in persistently recording all transactions in a decentralized manner, which bolsters user confidence in the decisions made by AI models [39].
3. **Collaborative decision-making:** Many applications benefit from decentralized decision-making processes. AI models often make decisions through a polling mechanism, with the majority ruling [40]. An AI predictor can log its decision on the blockchain, ensuring transparency for all participants. This process is repeated for each predictor until a final decision is reached.
4. **Secure data sharing:** Blockchain promotes data sharing by offering transparency, reliability, and accountability. It controls who can access user data, when it can be accessed, and who is accessing it. This feature is particularly beneficial in healthcare, where it can aid patients with rare diseases by allowing doctors worldwide to access and learn from similar cases [41].

5. **Decentralized intelligence:** High-level AI decisions can be made in a decentralized environment, where multiple predictors or AI agents work collaboratively with the same dataset. For example, in cybersecurity, AI agents can collectively work to ensure comprehensive network security [42].
6. **Increased efficiency:** Business processes involving multiple stakeholders, such as individuals, government organizations, and businesses, often suffer from inefficiencies due to the complexity of multiparty involvement. BlockXAI facilitates rapid and automated verification of data or asset transfers among diverse stakeholders, enhancing overall efficiency [43].

1.7 Conclusion and Future Work

This chapter provides an overview of blockchain technology, explainable artificial intelligence (XAI), and the synergistic combination of these two technologies, along with their applications and benefits. BlockXAI emerges as a solution to address key issues in AI technology, such as trust, security, and explainability, while leveraging the strengths of blockchain, including trust, non-repudiation, high efficiency, and enhanced data security. It is particularly useful in applications involving privacy preservation, data sharing, decentralized intelligence, and ensuring the trustworthiness of AI decisions. This technology augments decentralized AI systems by utilizing a digital ledger, which enables XAI agents to achieve consensus within the network.

However, several emerging challenges guide future research directions. Most BlockXAI approaches rely on smart contracts to automate processes. These contracts, which can be written in various programming languages like Go and Java, are not immune to vulnerabilities or threats [45]. A breach in a smart contract could fundamentally undermine the integrity of the blockchain. Therefore, the development of secure smart contracts remains an ongoing research challenge. Additionally, the success of a blockchain system largely depends on the effective implementation of a consensus mechanism. Designing an optimal consensus protocol and selecting the best one from the available options pose significant challenges.

Another consideration is the scalability of blockchain, particularly in terms of transaction rate and data storage capacity. While it is impractical to store extensive training data directly on the blockchain, decentralized data storage systems that support blockchain technology can be a viable solution. Furthermore, implementing an effective incentive mechanism on the blockchain is crucial. It encourages coordination among participants in computing AI models [8], with the nature of the incentive varying depending on the developers' choices. Lastly, the typically low throughput rate of public blockchains is a limitation, especially for real-time, sensitive AI applications. Therefore, enhancing the performance and throughput of blockchain systems is essential.

References

[1] PK, Fathima A. P K. (1984) "Artificial Intelligence-The next Revolution", in Dr. J. Karthikeyan, Dr. Ting Su Hie, Dr. Ng Yu Jin (eds) Learning Outcomes of Classroom Research, Tamil Nadu, L Ordine Nuovo Publication.

[2] Zhang, Song, Liu, Yin, Wang, & Dapeng Lan (2021) "Recent advances in blockchain and artificial intelligence integration: feasibility analysis, research issues, applications, challenges, and future work." *Security and Communication Networks* volume 2021: 1-15

[3] Gupta, Zhang, and Fei Wang (2016) "Model accuracy and runtime tradeoff in distributed deep learning: A systematic study" In 2016 IEEE 16th International Conference on Data Mining (ICDM):pp. 171-180.

[4] Shin, Donghee. (2021) "The effects of explainability and causability on perception, trust, and acceptance: Implications for explainable AI" *International Journal of Human-Computer Studies*146, 102551.

[5] Banerjee, Lee, and Choo, K. K. Raymond. (2018) "A blockchain future for internet of things security: a position paper." *Digital Communications and Networks*4(3): 149-160.

[6] Nassar, Salah, ur Rehman, and Svetinovic Davor (2020). "Blockchain for explainable and trustworthy artificial intelligence." *Wiley Interdisciplinary Reviews: Data Mining and Knowledge Discovery*10(1):e1340.

[7] Nakamoto, Satoshi (2008) "Bitcoin: A peer-to-peer electronic cash system." *Decentralized Business Review* 21260.

[8] Mohanta, Panda, and D Jena (2018) "An overview of smart contract and use cases in blockchain technology." *International conference on computing, communication and networking technologies (ICCCNT)*:1-4.

[9] Wang, Yuan, Wang, Li, Qin, and Wang F. Yue (2018) "An overview of smart contract: architecture, applications, and future trends." *IEEE Intelligent Vehicles Symposium (IV)*:108-113

[10] Androulaki, Barger, Bortnikov, Cachin, Christidis, De Caro, and Jason Yellick (2018) "Hyperledger fabric: a distributed operating system for permissioned blockchains." *Proceedings of the thirteenth EuroSys conference:1*-15.

[11] Zhang, Wu, and, Xinyi Wang (2020) "Overview of Blockchain consensus mechanism." *International Conference on Big Data Engineering:* 7-12.

[12] Li, Jiang, Chen, Luo and, Qiaoyan Wen (2020) "A survey on the security of blockchain systems." *Future Generation Computer Systems*107: 841-853.

[13] King and Scott Nadal (2012) "Ppcoin: Peer-to-peer crypto-currency with proof-of-stake." *Self-published paper.*

[14] Lin and Tzu-Chun Liao (2017) "A survey of blockchain security issues and challenges" *Int. J. Netw. Secur19(*5): 653-659.

[15] Juan Benet (2014) "Ipfs-content addressed, versioned, p2p file system." *arXiv preprint arXiv*:1407.3561.

[16] Wilkinson, Boshevski, Brandoff, and Vitalik Buterin (2014) "Storj a peer-to-peer cloud storage network." *CiteSeer*

[17] Ascigil, Reñé, Król, Pavlou, Zhang, Hasegawa, and Kentaro Kita (2019) "Towards peer-to-peer content retrieval markets: Enhancing IPFS with ICN." *ACM Conference on Information-Centric Networking*: 78-88.

[18] McConaghy, Marques, Müller, Jonghe, McConaghy, McMullen, and Granzotto (2016). "Bigchaindb: a scalable blockchain database." *White paper, BigChainDB.*

[19] Hartman, Murdock, and T. Spalink (1999) "The Swarm scalable storage system." *9th IEEE International Conference on Distributed Computing Systems(Cat. No. 99CB37003:* pp. 74-81).

[20] Mehrabi, Morstatter, Saxena, Lerman, and Aram Galstyan (2021) "A survey on bias and fairness in machine learning." *ACM Computing Surveys (CSUR)*54(6):1-35.

[21] Singh, A., & Singh, K. K. (2023). Blockchain with Federated Learning for Secure Healthcare Applications. Blockchain and Deep Learning for Smart Healthcare, 35-44.

[22] Tim Miller (2019) "Explanation in artificial intelligence: Insights from the social sciences" *Artificial intelligence*267:1-38.

[23] Babu, E. S., Dadi, A. K., Singh, K. K., Nayak, S. R., Bhoi, A. K., & Singh, A. (2022). A distributed identity-based authentication scheme for internet of things devices using permissioned blockchain system. Expert Systems, 39(10), e12941.

[24] Wang, Yang, Abdul, and Brian Y. Lim (2019) "Designing theory-driven user-centric explainable AI." Conference on Human Factors in Computing Systems:1-15.

[25] Brian Lim (2011) "Improving understanding, trust, and control with intelligibility in context-aware applications." *Human-Computer Interaction Institute, Carnegie Mellon University*.

[26] Nebula AI Team (2018) "NEBULA AI (NBAI)—Decentralized Ai Blockchain Whitepaper."

[27] Mohapatra, D., Bhoi, S. K., Jena, K. K., Nayak, S. R., & Singh, A. (2022). A blockchain security scheme to support fog-based internet of things. *Microprocessors and Microsystems*, *89*, 104455.

[28] Hasan and Khaled Salah (2018). "Blockchain-based proof of delivery of physical assets with single and multiple transporters." *IEEE Access*6:46781-46793.

[29] Marwala and Bo Xing (2018) "Blockchain and artificial intelligence." *ArXiv preprint arXiv*:1802.04451.

[30] Salah, Rehman, Nizamuddin and Ala AI-Fuqaha (2019) "Blockchain for AI: Review and Open Research Challenges." *IEEE Access*7:10127-10149.

[31] Gupta, R., Rakhra, A., & Singh, A. (2021). Internet of Things security using AI and blockchain. Machine learning approaches for convergence of IoT and blockchain, 57-91.

[32] Fu., Dong, and Ju, Yanbing (2020). "An intelligent electric vehicle charging system for new energy companies based on consortium blockchain." *Journal of Cleaner Production*261, 121219.

[33] Mao, Wang, Hao., and Li, Haisheng. (2018) "Credit evaluation system based on blockchain for multiple stakeholders in the food supply chain." *International journal of environmental research and public health*15(8), 1627.

[34] Pawlak., and Poniszewska-Marańda, Aneta (2019). "Blockchain e-voting system with the use of intelligent agent approach." In *Proceedings of the 17th International Conference on Advances in Mobile Computing & Multimedia* (pp. 145-154).

[35] Xu., Wang, and Guo, Mingayi (2017). "Intelligent resource management in blockchain-based cloud datacenters." *IEEE Cloud Computing*4(6): 50-59.

[36] Liu, Chen, Pei., Maharjan, and Zhang, Yan (2021). "Vehicular edge computing and networking: A survey." *Mobile networks and applications*26(3): 1145-1168.

[37] McGhin, Choo, Liu, and He, Debiao. (2019) "Blockchain in healthcare applications: Research challenges and opportunities." *Journal of Network and Computer Applications*135: 62-75.

[38] B Marr, B. (2018). Blockchain and the internet of things: 4 important benefits of combining these two mega trends. Forbes.

[39] Campbell, David (2018). "Combining AI and blockchain to push frontiers in healthcare." URL http://www.macadamian.com/2018/03/16/combining-ai-andblockchain-in-healthcare/, vol. online.

[40] Strobel, V., Castelló Ferrer, E., & Dorigo, Marco (2018) "Managing byzantine robots via blockchain technology in a swarm robotics collective decision making scenario." MIT Library (available at https://dspace.mit.edu/bitstream/handle/1721.1/115883/AAMAS2018.pdf?sequence=1&isAllowed=y)

[41] Oyelade, O. N., and Absalom E. Ezugwu (2020) "A case-based reasoning framework for early detection and diagnosis of novel coronavirus." *Informatics in Medicine Unlocked*20(100395)

[42] Brambilla, M., Ferrante, E., Birattari, M., and Dorigo, Marco (2013). "Swarm robotics: a review from the swarm engineering perspective" *Swarm Intelligence*7(1): 1-41.

[43] Magazzeni, McBurney, and Nash, William (2017) "Validation and Verification of Smart Contracts: A Research Agenda," *Computer*50(9): 50-57

[44] Zhuang, Y., Liu, Z., Qian, P., Liu, Q., Wang, X., and He, , Qinming. (2020) "Smart Contract Vulnerability Detection using Graph Neural Network" In *Proceedings of the Twenty-Ninth International Joint Conference on Artificial Intelligence (IJCAI-20)*:3283-3290.

2

XAI decision MODELS: Programming models for decentralized BlockXAI

Akansha Singh[1] **and Vijayalakshmi Saravanan**[2]

[1]School of CSET, Bennett University, India
[2]University of South Dakota, Department of Computer Science, USA

Abstract

This chapter presents an overview of the idea of explainable artificial intelligence (XAI) decision models and discusses the significance of these models in decentralized systems such as BlockXAI. This chapter investigates a variety of different programming models that can be utilized to construct XAI decision models. These models include decision trees, neural networks, and rule-based systems, among others. The advantages and disadvantages of each model, as well as their suitability for various kinds of decentralized applications, are examined in this chapter. In addition, this chapter discusses the difficulties associated with putting XAI decision models into BlockXAI and offers some suggestions as to how those difficulties could be overcome. In order to improve user trust and acceptability, the chapter comes to a close by highlighting the importance of having XAI decision models in BlockXAI that are both transparent and interpretable.

Keywords: Blockchain, smart contracts, Decentralized, Explainable AI, XAI Models.

2.1 Introduction

Machine learning models that are designed to make judgments in a decentralized and distributed environment are referred to as "decentralized XAI" decision models. XAI stands for "explainable artificial intelligence."

The decision-making process for a decentralized XAI system is not concentrated in a single node or server like it is in a centralized system. Instead, it is disseminated throughout a number of nodes or devices, some of which might be located in different parts of the world. Providing increased transparency, accountability, and reliability in decision-making processes is the objective of decentralized XAI decision models [1]. This is especially important in highly sensitive and important domains such as the healthcare industry, the financial sector, and the security sector. Concerns regarding bias, fairness, and privacy in machine learning systems can be helped to be addressed by using decentralized XAI decision models. These models can give a transparent and auditable record of the decision-making process, which can be helpful in addressing these concerns. XAI decision models that are decentralized can be implemented through a variety of implementation strategies, such as distributed ledger technology, federated learning, swarm intelligence, and multi-agent systems. These techniques make it possible to divide the decision-making process among numerous nodes, which can lead to improvements in scalability, fault tolerance, and resilience of the system [2].

Overall, decentralized XAI decision models offer a promising approach to building more transparent, accountable, and reliable machine learning systems that can be trusted to make critical decisions in complex, decentralized environments. This is because decentralized XAI decision models take advantage of the fact that machine learning algorithms are becoming increasingly complex.

2.2 XAI Decision Models

XAI decision models, also known as explainable artificial intelligence decision models, are forms of machine learning models that are intended to offer justifications for the choices they make. Unlike traditional machine learning models, also known as "black boxes," which produce outputs without providing any insight into the decision-making process, XAI decision models aim to provide a transparent and interpretable understanding of how decisions are made. This is in contrast to traditional machine learning models, which produce outputs without providing any insight into the decision-making process [3].

XAI decision models are particularly important in applications such as healthcare, finance, and security, where the repercussions of a decision can have a significant impact on human lives. XAI decision models are also important in other contexts where decisions can have a significant impact on human lives [4]. It is vital to have a comprehensive knowledge of how the machine learning model arrived at a choice in these fields in order to ensure that the result is fair, unbiased, and based on relevant variables. Only then can one be convinced that the decision was reached by the machine learning model.

The construction of XAI decision models can be accomplished through the application of a variety of methods [5]. The utilization of decision trees, which offer a framework that is both understandable and straightforward, is one strategy that can be taken. Using model-agnostic approaches is another strategy that can be taken. Model-agnostic methods, such as LIME (local interpretable model-agnostic explanations) and SHAP (SHapley Additive exPlanations), can provide explanations for any black-box machine learning model.

It is also possible to integrate XAI decision models with other methods in order to produce decision-making that is both more robust and more visible. For instance, in a decentralized XAI system, decision-making can be dispersed among numerous nodes or devices [6]. This not only improves scalability and fault tolerance, but it also provides a record of the decision-making process that is transparent and auditable.

Overall, XAI decision models are an important area of research in machine learning because they can help to improve transparency, accountability, and reliability in decision-making processes, while also ensuring that the decisions are based on relevant factors and are free from bias. This makes them an important area of research in the field of machine learning.

2.3 Types of XAI Decision Models

There are many different kinds of XAI decision models, and each one has its own set of advantages and disadvantages. Here are several examples:

- **Rule-based models:** In order to make a determination, these models follow a predetermined set of if–then guidelines. They are simple to comprehend and interpret, but they are not necessarily suited for the process of making complex decisions [7].
- Models known as decision tree models arrive at their conclusions via a decision-making process organized in a hierarchical fashion [8]. They are simple to view and analyze, but it is possible that massive datasets are not the best fit for them.

- **Bayesian networks:** These models base their judgments on the principles of probability theory. They come in handy in circumstances in which there is either uncertainty or a lack of complete information [9].
- **Fuzzy logic models:** In order to reach a conclusion, these models make use of fuzzy set data. They come in use in circumstances in which the data are either ambiguous or imprecise in some way.
- **Neural network models:** These models make decisions with the assistance of artificial neural networks. They are highly effective and are able to handle the making of complex decisions, but they might be challenging to interpret.
- Model-agnostic methods are those that can provide explanations for any black-box machine learning model. Some examples of model-agnostic methods include LIME and SHAP. They come in handy in situations in which the underlying model cannot be interpreted.
- Hybrid models are those that integrate two or more of the aforementioned methodologies to produce decision-making that is more reliable and accurate.

In general, the selection of a XAI decision model is determined by the particular application in question as well as the balance that must be struck between interpretability and complexity.

2.4 Centralized and Decentralized XAI Decision Models

Centralized XAI decision models are those in which the process of decision-making is centralized and managed by a single entity, such as a single server or a cloud-based system. This type of model is also known as a centralized model. With these types of models, the data are gathered and analyzed at a centralized location, and the decision is then made based on the data [10]. Because the decision-making process cannot be easily audited, these models, although they may provide efficient decision-making, may also be prone to concerns such as prejudice and lack of transparency.

On the other hand, decentralized XAI decision models disperse the decision-making process among numerous nodes or devices. Each node contributes to the decision-making process based on its own local data, and each device uses its own data to contribute to the decision-making process [11]. Because each node can give an auditable record of its own decision-making process, these models are supposed to provide better openness and accountability than other models now in use. Because the decision-making process is dispersed across numerous nodes, decentralized XAI models can also provide more scalability and fault tolerance than their centralized counterparts.

XAI decision models that are decentralized can be implemented through a variety of implementation methods, including federated learning, blockchain, and peer-to-peer networks, amongst others. By using federated learning, the data is kept on the local devices, and only the most recent modifications are sent to the centralized server [12]. In models that are based on blockchain technology, the process of making decisions is decentralized across numerous nodes in a way that is both secure and transparent. Each node in a peer-to-peer network contributes to the decision-making process based on its own local facts, and the ultimate conclusion is arrived at through the process of reaching consensus.

Generally, while centralized XAI decision models may provide efficient decision-making, decentralized XAI decision models may give higher transparency, accountability, scalability, and fault tolerance. Both types of models have their advantages. The choice of model is determined by the particular application being addressed as well as the balance between openness and privacy that is required.

2.5 Decentralized XAI Decision Models

A sort of machine learning model known as a decentralized XAI decision model is one that makes use of a distributed decision-making process. This means that the process of making decisions is spread out across numerous nodes or devices [13]. In these types of models, each node is responsible for contributing to the decision-making process based on its own collection of local facts, and the conclusion is arrived at through a process of reaching consensus.

When compared to centralized models, XAI decision-making frameworks that are decentralized are intended to give increased levels of openness, accountability, and fault tolerance. As a result of the distributed nature of the decision-making process, each node is able to offer an auditable record of the decision-making process that it employs, which enables more openness and accountability. In addition, because the data is stored locally, decentralized models can lessen the risk of both data breaches and invasions of personal privacy.

XAI decision models that are decentralized can be implemented through a variety of implementation methods, including federated learning, blockchain, and peer-to-peer networks, amongst others. By using federated learning, the data is kept on the local devices, and only the most recent modifications are sent to the centralized server. In models that are based on blockchain technology, the process of making decisions is decentralized across numerous nodes in a way that is both secure and transparent [14]. Each node in a peer-to-peer network contributes to the decision-making process based on its own local facts, and the ultimate conclusion is arrived at through the process of reaching consensus.

Decentralized XAI decision models are particularly beneficial in areas such as healthcare and finance, in which the data is spread across numerous sites. In these kinds of applications, decentralized models have the potential to offer improved levels of privacy and safety, in addition to facilitating higher levels of openness and responsibility in decision-making [15].

In general, decentralized XAI decision models are an important area of research in the field of machine learning. This is due to the fact that, in comparison to centralized models, decentralized models can provide greater transparency, accountability, and fault tolerance, all while addressing concerns related to privacy and lowering the risk of data breaches.

2.6 Categories of Decentralized XAI Decision Models

Depending on the particular implementation and the underlying technology that is being utilized, decentralized XAI decision models can be categorized in a variety of different ways. Here are several examples:

Federated learning is a method in which the data is stored on the users' local devices, and only the most recent updates are transmitted to a centralized server for the purpose of aggregate. Each device trains its own local model based on the data it collects, and then all of the models are integrated into one for the final judgment. Applications like healthcare, in which patient confidentiality is paramount, are good candidates for using federated learning.

Models that are based on the blockchain: Under this approach, the process of decision-making is decentralized across several nodes in a way that is both secure and transparent. Each node maintains its own independent copy of the blockchain, which serves both as a repository for the data and as the basis for the decentralized decision-making process. Applications such as finance, which place a premium on openness and accountability, are good candidates for using models built on blockchain technology.

Peer-to-peer networks: This strategy involves each node in the network contributing to the decision-making process based on its own local data, with the goal of arriving at a final conclusion by consensus. Applications such as cybersecurity, which place a premium on fault tolerance and decentralization, are good candidates for using peer-to-peer (P2P) networks.

Decision trees can also be distributed among numerous nodes, with each node being responsible for a certain section of the tree. This method is referred to as the distributed decision tree technique. The process of making decisions is decentralized, and the voting mechanism is used to determine which option should be chosen as the final option. Applications like picture identification, in which the data is spread out across a number of different places, are good candidates for using distributed decision trees. Decentralized artificial intelligence (AI) decision models can be broken down into numerous types, each of which has its own set of benefits and drawbacks. The appropriate model to use is determined by the particular use case as well as the balance that must be struck among efficiency, transparency, and personal privacy.

References

[1] E. Mbunge, B. Muchemwa, S. Jiyane and J. Batani, "Sensors and healthcare 5.0: transformative shift in virtual care through emerging digital health technologies", Global Health Journal, vol. 5, no. 4, pp. 169-177, 2021.

[2] C. C. Yang, "Explainable artificial intelligence for predictive modeling in healthcare", Journal of Healthcare Informatics Research, pp. 1-12, 2022.

[3] G. Minopoulos and K. E. Psannis, "Opportunities and challenges of tangible xr applications for 5g networks and beyond", IEEE Consumer Electronics Magazine, pp. 1-1, 2022.

[4] Pradhan, N., Singh, A. S., Sagar, S., Singh, A., & Elngar, A. A. (2021). Application of Machine Learning and IoT for Smart Cities *Machine Learning Approaches for Convergence of IoT and Blockchain*, 109-128.

[5] Khurana, R., Choudhary, M., Singh, A., & Singh, K. K. (2023). AIML-Based Blockchain Solutions for IoMT. *Blockchain and Deep Learning for Smart Healthcare*, 73-93.

[6] R. Di Pietro and S. Cresci, "Metaverse: Security and privacy issues", 2021 Third IEEE International Conference on Trust Privacy and Security in Intelligent Systems and Applications (TPS-ISA), pp. 281-288, 2021.

[7] J. Thomason, Metahealth - how will the metaverse change health care?, 2021.

[8] B. Mohanta, P. Das and S. Patnaik, "Healthcare 5.0: A paradigm shift in digital healthcare system using artificial intelligence iot and 5g communication", 2019 International Conference on Applied Machine Learning (ICAML), pp. 191-196, 2019.

[9] R. Gupta, S. Tanwar, S. Tyagi, N. Kumar, M. S. Obaidat and B. Sadoun, "Habits: Blockchain-based telesurgery framework for healthcare 4.0", 2019 International Conference on Computer Information and Telecommunication Systems (CITS) Beijing China, pp. 1-5, 2019.

[10] Saravanan, V., Aneja, I., Yang, H., Pillai, A. S., & Singh, A. (2021). Impact of big data in healthcare system—a quick look into electronic health record systems. In *Machine Learning and the Internet of Medical Things in Healthcare* (pp. 251-262). Academic Press.

[11] A. Adadi and M. Berrada, "Explainable ai for healthcare: From black box to interpretable models", Embedded Systems and Artificial Intelligence, pp. 327-337, 2020.

[12] Sharma, U., Ganapathi, A., Singh, A., & Singh, K. K. (2023). Blockchain in Healthcare: Use Cases. *Blockchain and Deep Learning for Smart Healthcare*, 147-169.

[13] A. Raza, K. P. Tran, L. Koehl and S. Li, "Designing ecg monitoring healthcare system with federated transfer learning and explainable ai", Knowledge-Based Systems, vol. 236, pp. 107763, 2022.

[14] R. Gupta, S. Tanwar, S. Tyagi and N. Kumar, "Tactile-internet-based telesurgery system for healthcare 4.0: An architecture research challenges and future directions", IEEE Network, vol. 33, no. 6, pp. 22-29, 2019.

[15] Mondal, S. S., Mandal, N., Singh, K. K., Singh, A., & Izonin, I. (2022). Edldr: An ensemble deep learning technique for detection and classification of diabetic retinopathy. *Diagnostics*, *13*(1), 124.

3

Blockchain-based frameworks for Explainable AI

Ivan Izonin[1], Krishna Kant Singh[2], and Akansha Singh[3]

[1]Department of Artificial Intelligence, Lviv Polytechnic National University, Ukraine
[2]Delhi Technical Campus, India
[3]School of CSET, Bennett University, India

Abstract

In this chapter, we look at how blockchain and explainable artificial intelligence (XAI) interact with one another. An introduction to blockchain and its possible uses in AI systems is provided to start this chapter. The need of explainability and openness in AI decision-making is then discussed, along with the notion of XAI. This chapter delves into the topic of using blockchain as a basis for building trustworthy and transparent XAI systems. This chapter also discusses the difficulties of developing XAI systems based on the blockchain and offers suggestions for overcoming these difficulties. Insights into how blockchain might be used to increase the openness and interpretability of AI systems, hence boosting their overall performance and social acceptability, are provided in this chapter.

Keywords: Blockchain, Explainable AI, Blockchain frameworks, AI Decison making.

3.1 Introduction

Explainable AI, also known as XAI, is a burgeoning subfield of AI research that aims to develop techniques and applications that will assist humans in comprehending the inner workings of AI systems and the conclusions they reach. Using frameworks that are based on blockchain technology is a strategy that holds a lot of promise for reaching XAI.

The blockchain is a distributed ledger system that enables record-keeping that is all at once safe, transparent, and unchangeable. Because it can ensure that the data used to train AI models is secure and auditable, it has the potential to create a decentralized and transparent platform for XAI. This is because it can prevent tampering with the data [1].

3.2 Blockchain-based XAI Frameworks

In general, frameworks that are based on blockchains present interesting potential for the research and development of XAI as shown in Figure 3.1. By capitalizing on the benefits offered by blockchain technology, we will be able to construct AI systems that are more open, safe, and decentralized, earning the confidence of all relevant parties.

a) **Transparency and auditability:** Artificial intelligence models and the procedures by which they make decisions can be transparently recorded and audited if blockchain technology is used. This indicates that the judgments made by an AI system may be linked back to the data and techniques that were utilized in the process of training the system [2]. This can serve to improve the level of trust that people have in AI systems and enable more informed decision-making.
b) **Data privacy and security:** Blockchain technology has the potential to be used to store and handle sensitive data, such as that required to train AI models, while also preserving data privacy privately and securely. This can assist to alleviate concerns about the misuse of data or breaches of data security, both of which have the potential to damage trust in AI systems.
c) **Decentralized governance:** Blockchain technology can make it possible to make decisions in AI systems without relying on a central authority. This demonstrates that the governance of an AI system can include participation from several different parties, and that decisions can be reached via agreement. This can be helpful in ensuring that the decision-making processes of AI systems are democratic, accountable to their users, and transparent [3].
d) **Incentive mechanisms:** Blockchain-based frameworks can be used to motivate individuals or organizations to provide data or participate in the governance of AI systems. This can be done through the usage of blockchain-based incentive mechanisms. Tokens, for instance, might

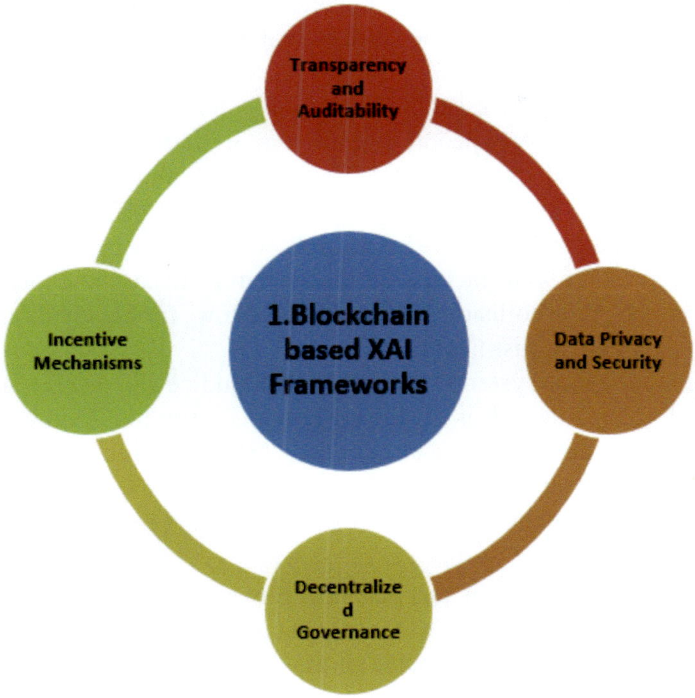

Figure 3.1 Blockchain-based XAI frameworks.

be used to offer an incentive to data suppliers or validators who work toward the creation of an AI system. This may assist to build an ecosystem for the development of AI that is more conducive to collaboration and openness.

3.3 AI Decision-making: Explainability and Openness

From smart personal assistants to autonomous vehicles, AI has permeated every aspect of modern life. As the prevalence of AI grows, so too is the importance of ensuring that its systems are open and easy to understand. Potential hazards and ethical considerations linked with AI systems necessitate explainability and openness in AI decision-making.

The trust issue is a crucial reason why AI decision-making should be accessible and explainable. Decisions made by AI systems, such as those used in finance or medicine, can have profound effects on people's daily lives [4]. Therefore, it is important for people to have faith in AI systems and to comprehend how the AI arrived at its conclusion. Lack of explainability undermines faith in AI systems because it makes it harder for people to comprehend and trust them.

Responsibility is another essential feature. When an AI system's choice has negative consequences for people or communities, it is important to know how the system reached its conclusion. Explainability and transparency in AI decision-making can aid in identifying and punishing offenders in such situations. For instance, if a self-driving car were to crash, it would be important to figure out what factors led the system to make the choice that ultimately resulted in the collision. Explainability allows for the identification of those at fault and their justification.

Artificial intelligence systems can be taught prejudice and discrimination if they are fed skewed data. It is challenging to recognize and remove these prejudices without transparency and openness [5]. The ability to discover and reduce prejudice and discrimination in AI systems depends on the transparency and explainability of their decision-making processes.

Human oversight: Because of the potential for error in AI systems, it is crucial to have human oversight in life-or-death circumstances. It is important for AI to be transparent and explainable so that humans can comprehend the reasoning behind its decisions and intervene when necessary [6].

Governments around the world are rapidly enacting legislation to regulate AI systems to make sure they are open and accountable. With AI decision-making that is both transparent and explainable, ethical norms can be enforced through legislation and regulation.

In conclusion, AI decision-making needs to be explainable and open to ensure that AI systems are trustworthy, transparent, and accountable. Without explanation, people lose faith in the system, and it becomes exceedingly difficult to determine who is at fault in emergency situations [7]. Further, governments can successfully control and legislate AI systems if they are explainable and open. This is because they can better uncover and eliminate prejudice and discrimination in these systems.

3.4 Challenges Faced

The goal of XAI systems is to make AI decisions understandable and accessible to humans. Blockchain is a secure and distributed ledger system that provides immutable data and promotes openness and auditability [8].

However, there are a number of challenges that must be overcome when building blockchain-based XAI systems.

- **Complexity:** Blockchain is a complicated technology, and XAI system integration will necessitate expert knowledge of both blockchain and XAI. Building blockchain-based XAI systems calls for expertise in areas such as machine learning, cryptography, and blockchain itself. Because of blockchain's inherent complexity, building and testing XAI systems can be time-consuming and expensive.
- **Limited storage and computational power:** Blockchain technology's limited storage and computational capacity makes it difficult to install sophisticated AI algorithms. Due to the high computational and data storage needs of machine learning models, implementing blockchain-based XAI systems can be difficult and expensive.
- **Data privacy:** Despite the importance of data privacy in AI decision-making, blockchain technology is built on the principle of transparency. It is important to strike a balance between openness and confidentiality while building blockchain-based XAI systems. Furthermore, once data is stored in a blockchain, it cannot be altered. However, there are times when data needs to be updated, which can be difficult to do without making mistakes [9].
- **Performance:** Blockchain's performance is constrained by the time it takes to process transactions. Artificial intelligence systems necessitate the rapid processing of massive datasets, which can be difficult to do in a blockchain setting. When blockchain-based XAI systems experience delays in processing transactions, it can cause delays in making decisions, which can lead to inefficiencies and poor performance.
- **Governance:** Blockchain's decentralized and autonomous nature makes it difficult to establish traditional administrative frameworks for XAI systems. To guarantee the responsible and ethical operation of XAI systems, governance mechanisms are necessary. The development of blockchain-based XAI systems is complicated by the technology's decentralized nature, which makes it hard to create governance mechanisms.

Developing XAI systems on a blockchain infrastructure faces many obstacles, including those related to data protection, performance, and governance. While blockchain's data immutability, transparency, and accountability are attractive, the technology also presents challenges to the implementation of AI algorithms [10]. In order to overcome these obstacles, specialists in AI and blockchain technology should work together to create XAI systems that are both scalable and efficient.

3.5 Suggestions

Experts in both AI and blockchain technologies will need to work together to overcome the challenges of establishing blockchain-based XAI systems. Some solutions to these problems are listed below.

- **Collaboration:** Experts in both AI and blockchain technology must work together to create XAI systems that are both scalable and efficient. Experts in artificial intelligence can contribute understanding of AI algorithm development, while blockchain specialists can offer insight into distributed ledgers, smart contracts, and cryptography. Working together, these professionals can find solutions to the problems that arise when combining AI and blockchain [11].
- **Optimization:** Algorithm optimization overcoming obstacles in establishing blockchain-based XAI systems relies heavily on optimizing AI algorithms. In order to mitigate the effects of limited storage and computational power, researchers can build algorithms that are both computationally efficient and demand little space for storing. The efficiency of blockchain-based XAI systems can also be enhanced by optimizing algorithms.
- **Privacy-preserving techniques:** The development of blockchain-based XAI systems necessitates the use of privacy-preserving techniques. Data privacy can be preserved and AI decision-making transparency achieved through the use of privacy-preserving approaches [12]. Data privacy can be preserved and transparent AI decision-making achieved through the use of methods like homomorphic encryption and secure multi-party computation.
- **Governance models:** For blockchain-based XAI systems to be run in an ethical and accountable manner, it is essential that governance models be developed that can function in a decentralized setting. Governance models can be developed by researchers to guarantee openness and responsibility in AI decision-making while still allowing for decentralized decision-making [13].
- **Use of off-chain data storage:** Off-chain data storage is used to compensate for blockchain's insufficient processing and storage capacity. The impact of the blockchain's limited storage and processing capacity

can be mitigated by storing data off-chain [14]. The efficiency of blockchain-based XAI systems can be enhanced by storing data off-chain.

In conclusion, a combined effort by AI and blockchain specialists is necessary to overcome the challenges of establishing blockchain-based XAI systems. Possible solutions to these problems include algorithm optimization, privacy-preserving approaches, the creation of governance mechanisms, and the utilization of off-chain data storage [15]. Transparent and responsible AI decision-making can be achieved through the combined efforts of professionals working on blockchain-based XAI systems that are both scalable and efficient.

References

[1] Mohapatra, D., Bhoi, S. K., Jena, K. K., Nayak, S. R., & Singh, A. (2022). A blockchain security scheme to support fog-based internet of things. *Microprocessors and Microsystems*, *89*, 104455.

[2] D. Berman, A. Buczak, J. Chavis and C. Corbett, "A survey of deep learning methods for cyber security", *Information*, vol. 10, no. 4, pp. 122, Apr. 2019.

[3] Gupta, R., Rakhra, A., & Singh, A. (2021). Internet of Things security using AI and blockchain. *Machine learning approaches for convergence of IoT and blockchain*, 57-91.

[4] Khurana, R., Choudhary, M., Singh, A., & Singh, K. K. (2023) Energy-Efficient Fog-Assisted System for Monitoring Diabetic Patients with Cardiovascular Disease. Blockchain and Deep Learning for Smart Healthcare, 323-352.

[5] Cyberattack Disrupts Unemployment Benefits in Some States, Jul. 2022, [online] Available: https://www.washingtonpost.com/politics/cyberattack-disruptsunemployment-benefits-in-some-states/2022/06/30/8f8fe138-f88a-11ec-81db-ac07a394a86b_story.html.

[6] Gaba, S., Khan, H., Almalki, K. J., Jabbari, A., Budhiraja, I., Kumar, V., ... & Abouhawwash, M. (2023). Holochain: An Agent-Centric Distributed Hash Table Security in Smart IoT Applications. IEEE Access.

[7] D. Gumusbas, T. Yldrm, A. Genovese and F. Scotti, "A comprehensive survey of databases and deep learning methods for cybersecurity and intrusion detection systems", *IEEE Syst. J.*, vol. 15, no. 2, pp. 1717-1731, Jun. 2021.

[8] S. Zeadally, E. Adi, Z. Baig and I. A. Khan, "Harnessing artificial intelligence capabilities to improve cybersecurity", *IEEE Access*, vol. 8, pp. 23817-23837, 2020.
[9] S. M. Mathews, "Explainable artificial intelligence applications in NLP biomedical and malware classification: A literature review" in Intelligent Computing, Cham, Switzerland:Springer, pp. 1269-1292, 2019.
[10] M. Sahakyan, Z. Aung and T. Rahwan, "Explainable artificial intelligence for tabular data: A survey", *IEEE Access*, vol. 9, pp. 135392-135422, 2021, [online] Available: https://ieeexplore.ieee.org/document/9551946.
[11] B. Goodman and S. Flaxman, "European union regulations on algorithmic decision-making and a 'right to explanation", *AI Mag.*, vol. 38, no. 3, pp. 50-57, Oct. 2017.
[12] M. Nazar, M. M. Alam, E. Yafi and M. M. Su'ud, "A systematic review of human–computer interaction and explainable artificial intelligence in healthcare with artificial intelligence techniques", *IEEE Access*, vol. 9, pp. 153316-153348, 2021.
[13] H. Jiang, J. Nagra and P. Ahammad, "SoK: Applying machine learning in security—A survey", *arXiv:1611.03186*, Nov. 2016.
[14] L. Buczak and E. Guven, "A survey of data mining and machine learning methods for cyber security intrusion detection", *IEEE Commun. Surveys Tuts.*, vol. 18, no. 2, pp. 1153-1176, Oct. 2016.
[15] D. Kwon, H. Kim, J. Kim, S. C. Suh, I. Kim and K. J. Kim, "A survey of deep learning-based network anomaly detection", *Cluster Comput.*, vol. 22, pp. 949-961, Jan. 2019.

4

Explainable Artificial Intelligence (XAI): Notions of Explainability and its Integration with Blockchain

Raj Gaurang Tiwari[1] and Mohammad Husain[2]

[1]Chitkara University Institute of Engineering and Technology, Chitkara University, India
[2]Faculty of Computer and Information Systems Islamic University of Madinah, Saudi Arabia
Email: rajgaurang@chitkara.edu.in; mohd.husain90@gmail.com

Abstract

Machine intelligence is referred to as "artificial intelligence." Computational intelligence is another name for it. An intelligent agent is a technology that acts in such a manner as to accomplish a pre-determined objective, while also observing its surroundings and learning from its experiences. This is the focus of AI research. Despite machine learning's enormous success, its flaws have grown increasingly apparent as a result of this success. Medical and government advancements in recent years have highlighted the fundamental issues with confidence in these and other key domains. It is a major shortcoming of AI systems because they cannot explain why they made certain decisions. AI systems' increasing independence and ability to operate on their own means that the decisions and actions they take are becoming more difficult for humans to follow and comprehend. But unless convincing explanations are provided, AI applications will not be able to achieve their potential. To produce machine learning approaches that yield more understandable models without compromising prediction accuracy, the explainable AI (XAI) was born. XAI aims to help humans trust, understand, and manage AI systems. XAI research is reviewed in this chapter, although the emphasis is on artificial intelligence (AI), which is closely related to XAI. By using blockchain, smart contracts, trusted oracles, and decentralized storage, we also shed light on more reliable XAI.

Keywords: Explainable Artificial Intelligence (XAI), Machine Learning, Artificial Intelligence (AI), Interpretability, Explainability.

4.1 Introduction

AI is machine intelligence. Computational intelligence is another term. AI involves the study of intelligent behavior and intelligent agents, technologies that operate to ideally attain a planned goal while observing the surrounding (data) environment and learning from the observations [1].

AI classification can be grouped by how close the AI can emulate human thinking. Thus, we can differentiate how the system compares to humans in terms of versatility and performance. Based on these criteria, we classify AI systems on their similarity to the human mind and their ability to "think." This ability could be classified into four basic categories: reactive machines, limited memory machines, a theory of mind [2], and self-aware AI [3].

Reactive: There is no memory in a reactive machine; it merely reacts to various stimuli. "Simplest" is an understatement for this sort of AI system. There is no memory in these devices, and they are quite restricted in what they can do. They are often preprogrammed to respond solely to a small subset of possible inputs. When it comes to chess, DeepBlue [1] is an example of reactive machine AI.

Limited memory machines: Capacity is limited. Unlike reactive systems, machines employ memory to enhance their reactions. These systems can enhance their judgments by using the knowledge they have gleaned from the past. Every single current application falls within this category (e.g., chatbot or self-driving car).

Theory of mind: Theory of mind-level AI will be able to grasp the needs, feelings, beliefs, and mental processes of the beings it is dealing with. It is still an idea or a work in progress. Human brains may be described in many ways; thus, researchers need to design a system that recognizes persons as individuals [2].

Self-aware AI: AI development has yet to reach this last step, which is now simply a theoretical concept. Self-aware AI has gained self-awareness because it is so close to the human brain [3].

As a result of its importance, machine learning (ML) emerged as a separate topic within artificial intelligence (AI). Artificial intelligence was no longer the primary goal of machine learning; instead, it was addressing real-world issues. It shifted its focus away from symbolic techniques and toward statistical and probability theory tools and models.

While deep neural networks have been around for a long time, Le et al. [4] revisited them in 2011 and built their first unsupervised network. Because of a lack of computer power and data, deep neural network theory was originally discarded by researchers in the 1990s. The "boom" in deep learning was sparked by this study.

4.2 Machine Learning and Artificial Neural Networks

In modern parlance, "AI" may be replaced by "ML." Figure 4.1 shows that machine learning is a subset of AI. Without machine learning, AI can be constructed. AI systems that do not use machine learning are massive ruled-based systems that forecast outputs [5, 6].

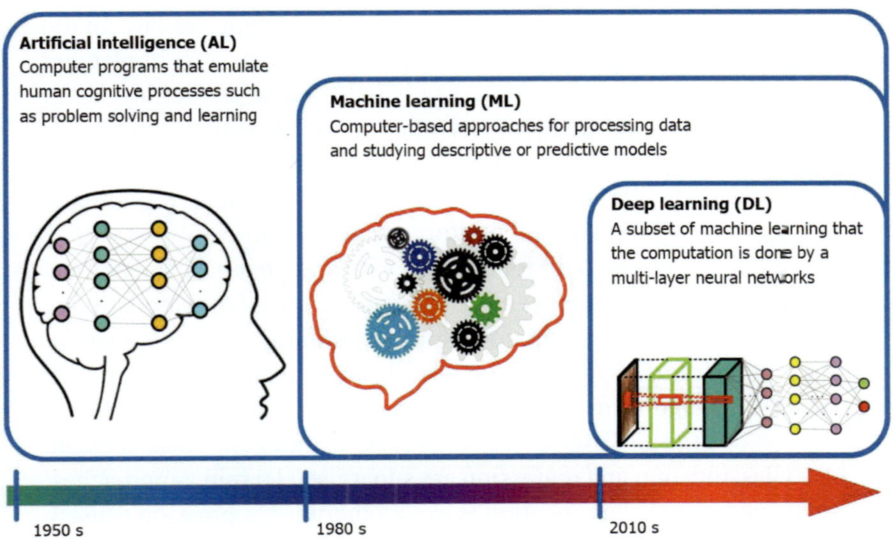

Figure 4.1 AI, machine learning, and deep learning timeline.

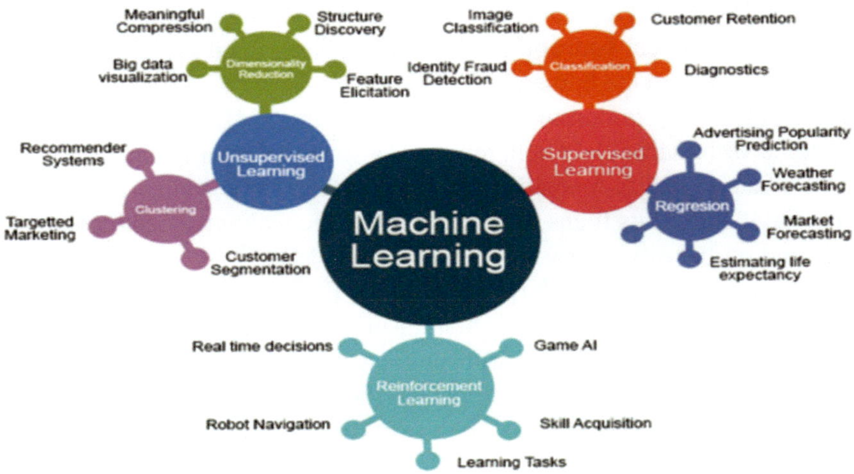

Figure 4.2 Illustration of machine learning categories.

Artificial intelligence is becoming more dependent on machine learning. By giving computers real-world data, machine learning explains the science of helping computers learn and improve their learning on their own. The term "learning" in this context refers to an individual's ability to enhance their performance in a certain activity via the use of statistical approaches. In the subject of machine learning, researchers investigate, design, and implement algorithms that can not only learn from data but also make predictions about it. As a result, ML is useful in situations when rules-based programming is unable to provide a solution. Email filtering is a good illustration of this.

Figure 4.2 depicts the three types of machine learning algorithms: (semi)supervised, unsupervised, and reinforcement learning.

Mathematical models are built for datasets with known inputs and outputs in semi-supervised or supervised learning techniques. It is necessary to divide the raw data into two sections: one for algorithm training, and the other for algorithm testing. Examples of input and intended output are included in each training example. Some semi-supervised learning training situations do not provide desired outcomes. Task-driven supervised learning algorithms create predictions. Classification and regression are supervised learning approaches. If the outputs are limited, classification is used; otherwise, regression is used. (Semi)supervised, unsupervised, and reinforcement learning are shown in Figure 4.2.

Unsupervised learning methods employ input-only data to find patterns and correlations. The algorithm groups or clusters input points based on similarities between the data and the taught structure. Cluster analysis is a kind of unsupervised learning. PCA separates correlated variables into linearly uncorrelated ones. The method shows how populations interact, for example.

Reinforcement learning algorithms develop via feedback. Software agents are continually looking for methods to maximize their condition, even if the return is short-term. Issue environments are commonly described using a discrete and stochastic Markov decision process.

Biological neural networks, such as those found in human brains, were the inspiration for artificial neural networks (ANNs), also known as neural networks (NNs). Artificial systems, like people, pick up new skills by observing and copying what they see around them. In the algorithms, there are no task-specific rules. Algorithms based on artificial neural networks (ANNs) have the potential to detect handwritten figures in images. There may be a large collection of handwritten figures, as well as textual representations of the figures. ANNs are frequently employed in artificial intelligence applications.

Figure 4.3 depicts an ANN system. It is a collection of artificial neurons, each of which communicates with the others through a link called an edge. It is possible to change the weight of the signal to represent the significance

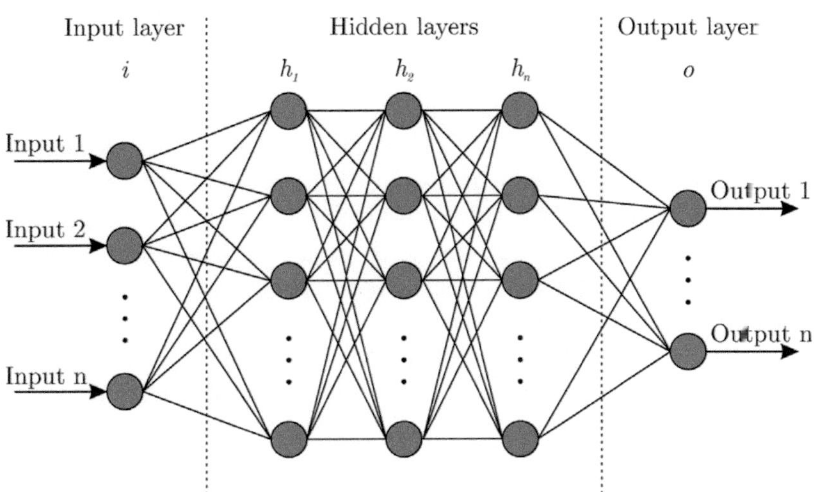

Figure 4.3 Illustration of machine learning categories.An illustration of a synthetic neural network (ANN). Neurons are organized into three types of layers based on where they are in the processing chain: input, hidden, and output.

of the connection. In ANN implementations, a signal is often a real number. Each neuron communicates with and receives information from several other neurons, creating many connections.

There are frequently several layers of neurons in the network. The activity begins at the input layer and travels through (many) hidden levels until it reaches the output layer, which is considered the final layer of ANN. Each neuron transforms the information it receives in a layer-specific manner.

4.3 Need of XAI

The value of XAI and its ability to explain itself was recognized for decades [7]. The success of machine learning has opened numerous avenues for the use of artificial intelligence, but it has also shown its flaws. The trust issue has been highlighted by recent advancements in vital domains such as health and governance. The inability of AI systems to explain why a certain choice was taken is one of its most significant drawbacks [8]. Even if there is a lengthy history and a large amount of effort, it is not straightforward to come up with solid answers. AI systems are becoming more complicated and difficult to comprehend because of their reliance on millions of factors and because the models themselves are responding and deciding more and more autonomously [9]. Most often the smartness and transparency of AI are contradictory. But unless convincing explanations are provided, AI applications will not be able to achieve their potential.

For practical purposes, AI systems are designed to utilize training data to identify the most effective model. Consider, for example, maximizing the accuracy of a tumor's benign or cancerous status using the available data. AI may learn beneficial, generalizable rules like "benign tumors grow slower," but it may also learn false rules. This is particularly true if the training data utilized has any incorrect connections, such as a link between a tumor's diagnosis and its ID number in the data. The consequences may be quite severe if people were to learn these fictitious principles and then apply them to actual events in the world. If AI systems could explain their conclusions, these kinds of erroneous rules would be easier to discover and the model itself would be more trustworthy.

If there is no need for interpretation, the subject of whether or not it is necessary remains up to debate. In certain academic circles, interpretability is a matter of context. No human interpretation is required since the output of machine learning is not impacted by it, it has been thoroughly tested, and negative results will not have a significant impact. This assertion, on the other hand, ignores the role that interpretation plays in building general confidence in the system.

Doshi-Velez and Kim [10], on the other hand, consider the requirement of interpretation as being linked to a fully formalized issue. An example of incompleteness in this context would be a lack of scientific understanding about the subject under investigation or an inability to test the system to its fullest potential. If the formalization of the issue is not full, more investigation is required to fill up the gaps.

There are many reasons why explainable AI is required, not only incomplete problem formalization. In the beginning, a system's explanations might be utilized as a kind of validation. A trained AI system may come to incorrect conclusions if the dataset is skewed, which may be readily seen if the rationale is made public. The second need for improving AI is to be aware of and comprehend the system's flaws. It is difficult to find the flaws in black boxes. As the last point, the explanations may provide fresh insights. AI systems may discover new connections and deliver fresh insights since they are taught using vast datasets that are unavailable to humans. In the end, it is now a matter of law as much as explanations. The EU's new legislation specifies that everyone has the right to an explanation, which implies that AI-driven choices must also offer an explanation.

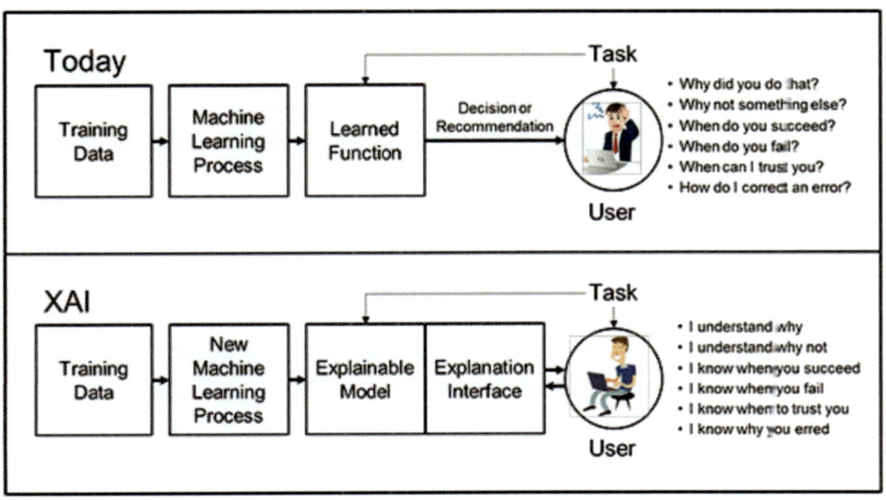

Figure 4.4 The key goals of the XAI program [9].

A user must be able to understand and accept the model's output to be pleased with it. Analyzing a prediction's explanations allows a user to evaluate the model and determine whether or not it is logical or correct [11].

Research organizations and funding agencies alike have recognized the urgent need for explanation and interpretation. The explainable AI (XAI) initiative, for example, was founded by DARPA to develop machine learning approaches that yield more explainable models without compromising prediction accuracy and that also enable humans to comprehend, trust, and manage new AI tools like XAI [9].

Figure 4.4 summarizes the main objectives of the XAI initiative. The top panel depicts how current AI apps are constructed and how difficult it is for the user and the computer to engage with one another. Explanations assist users in better grasping the AI system shown in the bottom panel of the illustration. Using the program's output, the user may deduce when and where the program is operating correctly, and where mistakes could occur [9].

4.4 Explainable Artificial Intelligence

Because in the past, AI algorithms were easily understandable, there was no need to explain the decisions made by these systems at first. Even the simplest deep neural network is difficult to understand today because of the complexity of today's algorithms. As these systems became more prevalent in our day-to-day lives, the need for an XAI became more apparent. It was unfortunate that these systems were so complex that no one could understand or predict the outcome of the given input. Healthcare, law, defense, and finance were among the industries where people's lives depended on the outcomes of these systems.

Figure 4.5 depicts an initial taxonomy of XAI methods based on previous research.

AI's predictions may be explained using XAI's methods. However, just because it is called "XAI" does not imply we have to see it as an extension of AI; rather, it is an issue of human–agent interaction. As a result, XAI must be able to effectively communicate its conclusion. The way we explain this choice depends on who we are talking to (examples of the difference are in Figure 4.6).

The capacity to explain why a computerized system performed specific actions, made predictions, offered suggestions, or made judgment is a critical component of artificial intelligence. Explainability, interpretability, and trust

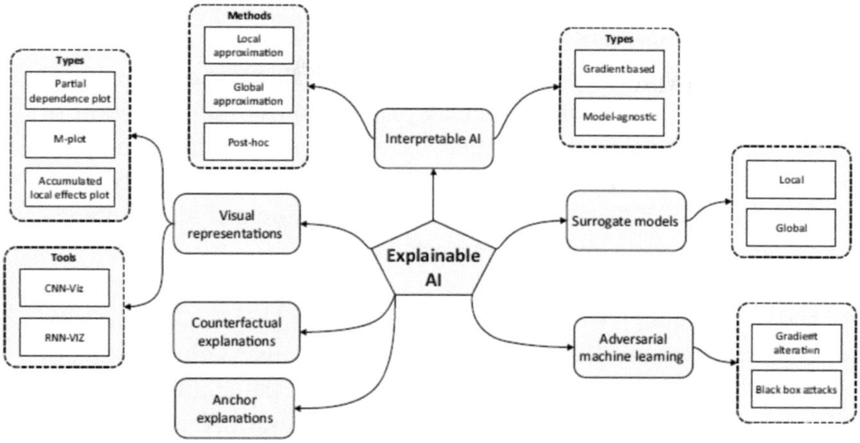

Figure 4.5 Taxonomy of XAI methods.

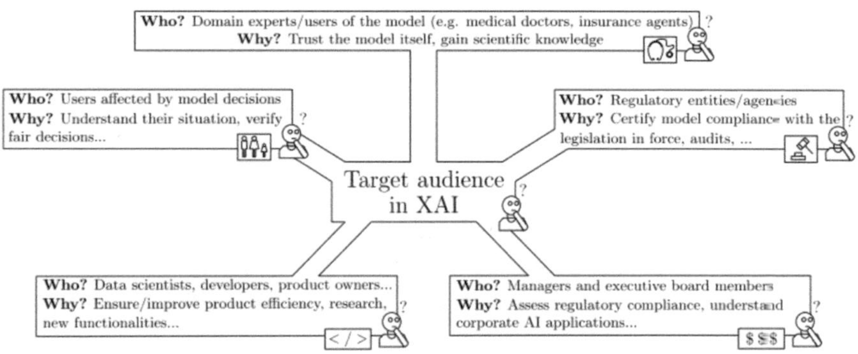

Figure 4.6 Explaining to the different target audiences.

are three major concepts used to define this ability in the literature. To confuse things further, academics use multiple names for the same concepts [8].

Sometimes in literature, "interpretability" and "explainability" are used interchangeably. Explainability and interpretability are used interchangeably in machine learning.

Interpretability: Understanding the reasoning behind a choice is referred to as interpretability.

Explainability: The capacity of a machine or deep learning system to be described in human terms is known as explainability. Being able to describe what is going on is what we mean when we talk about explainability.

Interpretability is defined by Doshi-Velez et al. [10] as the ability to explain or convey in terms that a human can comprehend. According to Samek et al. [12], interpretation is the transfer of abstract concepts into domains humans can understand, whereas explanation is the collection of elements in an interpretable domain that have led to a specific example's decision. There are two terms often used to describe interpretability: comprehensibility and transparency. According to Doilovic et al. [8], knowing the working logic of the model is important.

Taking into account the characteristics of the trustee (e.g., benevolence, integrity, and competence), trust is defined by Israelsen [13] as a psychological state in which a person voluntarily and securely becomes vulnerable to, or relies on, another person or institution (e.g., a human or artificially intelligent agent). According to Ribeiro et al. [14], trust may be described as a belief in a forecast or belief in a model. A user may either rely on a single forecast or the whole model to make choices, depending on their preferences. These ideas are intertwined, yet they also stand on their own. If a model's user can rely on just one prediction, then the model may be considered trustworthy.

4.5 XAI and Cognitive Sciences

How to explain XAI to humans has been studied extensively. An examination of how people construct and communicate explanations, as well as their use of cognitive biases and social expectations, ties into this issue.

There is little evidence that XAI research is based only on the researchers' own intuition and judgment regarding what constitutes a valid explanation. Despite the importance of understanding how people utilize and comprehend explanations, research frameworks from the social sciences are seldom used or referenced. AI model creators and specialists who are intimately familiar with AI models are not the best people to assess the level of explanations required by non-experts [15]. There has been much study into how to explain XAI to a human observer.

There is little evidence that XAI research is based only on the researchers' own intuition and judgment regarding what constitutes a valid explanation. Despite the importance of understanding how people utilize and comprehend explanations, research frameworks from the social sciences are seldom used or referenced. AI model designers and specialists who are intimately aware of AI models are not the best persons to determine the amount of explanations necessary for non-experts [15].

4.6 Explanations and Their Characteristics

The XAI community has also examined the essential aspects and explanations of XAI systems. Miller [15] says trustworthy anonymous systems should be able to:

1. generate judgments using an interpretability/explainability requirement;
2. elucidate choices to the human(explanation).

Explaining concepts is critical in the initial feature. The importance of being able to explain and interpret the results of an anonymous system's search for a solution cannot be overstated. The more comprehensible the answer is, the better it is. For example, if two distinct answers are roughly equivalent, the output is the one that is easier to explain. To further explain their decisions to people, anonymous systems should provide explanations.

For explanations, Miller [15]identifies four fundamental notions that are ubiquitous in explanations offered and accepted by humans.

1. Contrasts are used in explanations: People often wonder why a given occurrence transpired.
2. There is prejudice in the selection of explanations: People seldom anticipate a detailed cause list. They usually choose one or two factors.
3. Explanation does not need to provide probabilities: In explanations, probabilities are less effective than causes.
4. Sociological explanations: They are the opinions of the explainer in relation to the perspectives of the explainee.

Miller et al. [15] figured out why output A rather than output B was delivered. People, on the other hand, do not anticipate or care about hearing the whole explanation. There are several factors at play, but only a few stand out.

With regard to probabilities and causes, probabilities play a crucial part in many AI applications. As simple as it may seem, Miller [15] believes that offering causes rather than probabilities are a better method to explain the result for AI specialists. What type of explanations is most successful depends on the audience. According to Miller [15], people prefer to provide explanations that are based on the explainee's position.

It has been addressed in detail by Ribeiro, Singh, and Guestrin [14]. For them, specific traits are required in explanations. To begin with, the explanations must be understandable, and this is a function of both the intended audience and the nature of the issue being addressed. Users should be able to grasp and utilize explanations with ease. Second, the explanations must be accurate to the area being discussed. The model's behavior at the point of interest has to be properly described, although perfect fidelity of description cannot normally be asked for without diving into model specifics. Some information about the model's overall accuracy is also offered by a skilled explainer, who should be able to describe any model.

If a model is going to be helpful in the actual world, it must be tested at some point. Classification models, for example, are commonly evaluated on their ability to predict the outcome of a test dataset. Aside from being time-consuming, these tests might also provide inaccurate estimates of the model's accuracy since real-world data can vary greatly from the sample used for testing.

4.7 Different Types of Approaches to Explain and Interpret

Integrated (transparency-based) and post-hoc techniques [8] are the two basic types of explanation and interpretation methods. As a general rule, initiatives striving for transparency fall under the umbrella of integrated interpretability. It is possible that our justifications for how we came to a conclusion vary from the actual thought process that led to it.

Basic models like linear models and decision trees may explain themselves, but the audience must comprehend. Due to the model's simplicity and inflexibility, the technique only applies to basic systems like artificial neural networks.

The post-hoc method approaches the model as if it were some kind of enigmatic opaque container. In order to ensure that the information can be interpreted, it is all retrieved from what was previously learned [8]. Post-hoc approaches that treat a model as if it were a "black box" have no impact whatsoever on the performance of a model. Interpretability and/or explainability are the focus of post-hoc techniques. An approximation model is being sought in one technique, known as the transparent proxy model approach. Assche et al. [16] have utilized the approach to construct a single decision tree from many. Martens et al. [17] and Zhou et al. [18] utilized the method on SVMs and neural network ensembles.

Indicative approaches, such as visualization, may also give post-hoc explanations, although interpretability is less of a concern with these techniques. Instead, they focus on the model's features. In the past, several visualization approaches were put to use. Deconvolutionary networks were utilized by Zeiler et al. [19] to illustrate the layers of convolutional neural networks, while Karpathy et al. [20] employed visualization methods to explain recurrent neutral networks. To enhance the model's architecture and to identify cells in the text that handle long-range relationships, visualization methods were used in both instances. For example, neural networks or support vector machines may be shown using Cortez et al.'s [21] model-agnostic visualization technique based on a sensitivity analysis. There are visualization techniques that are not tied to any particular model, sometimes known as model-agnostic.

The explanatory strategy may merge these two groups. Then there are "third-category" techniques, such as hybrid ones.

4.8 Implemented XAI Applications

In 1970s, rule-based expert systems supplied explanations. Rule-based expert systems, the most basic kind of AI, employ if–then–else rules to communicate human knowledge about a topic. Other contexts include:

- Probabilistic decision-making Bayesian networks [22]
- Recommendation systems [23–25]
- Constraint programming [26]
- Context-aware systems [27]
- Markov decision processes [28]
- Case-based reasoning systems [29]
- Causal discovery [30]

4.8.1 Recommendation systems

An online service that provides individualized product suggestions is referred to as a recommendation system. Most XAI research has focused on Bayesian networks, rule-based expert systems, and recommendation systems [22–25]. There have been several research works on the sorts of justifications that individuals find most appealing in recommendation systems. Predictions may be explained by rating histograms [25]. Other factors included a user's past performance and the similarity of working items [23]. They discovered that the most useful explanations incorporate many distinct explanation types: it is preferable to explain a suggestion based on the user's choices, on comparable user choices, and on features rather than relying just on one form of explanation.

4.8.2 Constraint programming

Constraint solvers address combinatorial search problems with decision variables and restrictions. Explanation generation has been studied in non-ML systems. Jussien et al. [31] researched ways to convey conflict restrictions to the user, whereas Amilhastre et al. [32] offered methods to restore conflict situations to non-conflict ones.

4.8.3 Context-aware systems

"Context-aware" systems recognize their greater context. People are aware of their environment and react accordingly. Lim et al. [33] built a toolbox for context-aware systems that explains eight decision model types. It is crucial to know the program's input data. The application's outputs tell the user what type of outputs and capabilities they may anticipate from it (e.g., number of result options). To assist users understand what would happen if they entered a different set of input values, the "What If" explanation type shows them the previous or current output value. Explanations are a way to explain why a certain set of inputs led to a specific outcome. "Why Not" explains to the user why a certain outcome was not achieved. The term "How To" refers to instructions that show the user how to accomplish a certain goal. Last but not least, explanations of certainty convey the degree of certainty or uncertainty around the generated output value.

4.8.4 Markov decision processes (MDPs)

An approach to decision-making based on Markov decision processes (MDPs) is stochastic, in which both the decision-maker and randomness impact the outcome. Elizalde et al. [34] created an intelligent power plant helper. To help the user better comprehend the MDP planning system's proposed and produced commands, an explainable intelligent assistant explains the instructions. With its roots in power plant operations, the approach may be used in various areas where people's training or assistance is required.

Gattami et al. [28] describe Markov decision processes using two case studies and in a domain-independent manner.

In decision-support systems, explanations are critical. It was shown in a research of anesthesia medical support systems that users made fewer errors when explanations were offered and that they were more confident in their judgments and critical of the underlying model when explanations were supplied.

4.8.5 Explanations in machine learning

Visualization, prediction interpretation and justification, and interpretable models have all been investigated in the literature on machine learning.

Historically, machine learning professionals themselves have been the driving force behind the development of explanations, as they sought to assess whether or not the model was performing properly. As a preliminary step in determining if the model's predictions were accurate, a nomogram was employed to represent the data in question.

Recent research has focused on revealing the secret states of neural models, whereas nomograms have found use in a wide variety of contexts [35–37]. They produced a number of visualizations to illustrate the connections between the input and output data.

Prediction interpretation and justification seek to understand forecasts, for example, by emphasizing the contributions of various elements. Throughout the previous several decades, both model-specific and model-agnostic solutions have been presented to solve the problem. Model-agnostic interpretation techniques, on the other hand, are framework-dependent. Classification and natural language processing (NLP) have been suggested for model-agnostic techniques. Approximation strategies for complicated models have been developed, including global as well as local approaches [14] Model approximation involves building a simplified model that approximates the original's answer.

The precision of global approximations varies widely. Using an approximation that is closer to a place of interest, such as a local approximation, may provide superior accuracy. The LIME approach (local interpretable model-agnostic explanations) explains classifier predictions by fitting a basic model to the prediction's immediate surroundings. In image recognition, the explanatory model space is the pixels' vectors [14]. Peltola [38] has recently evolved LIME into a new methodology known as "KL-LIME," which combines LIME with predictive variable selection approaches.

In addition, the idea of interpretable models may be used as a means of focusing on explanation. Interpretable models, on the other hand, make an effort to be interpretable in and of themselves, rather than making an effort to explain black-box models. Models that may be interpreted include examples such as decision trees and rule-based models. In classification, for example, interpretable models have been developed. Bayesian techniques that integrate rule lists and probability distributions have also been studied.

4.8.6 Deep learning explanation approaches

For the purpose of better explaining the results of deep learning models, Samek et al. [12] introduced two approaches that are now in widespread use:

local regression planning (LRP) and sensitivity analysis. These methods were used to classify images and text documents and to recognize human actions.

In picture categorization, there has been a lot of development since it is so simple to see the explanations. Many methods have been developed to identify the most significant pixels in the AI system's output. Because these pixels have a substantial impact on system output, the goal is to identify them.

Analyzing for sensitivity is a common technique. Samek et al. [12] suggested sensitivity analysis and local regression planning (LRP) to elucidate deep learning model predictions (layer-wise relevance propagation). These algorithms were used to categorize photos, text, and human behaviors. The gradients of the input (parameters) are computed in this approach. Input characteristics that have the greatest impact on the final result are those that are most important. There is information on which pixels in a picture have the most impact on the choice from a sensitivity study conducted by Samek et al. [12].

It is also common to use layer-wise relevance propagation (LRP), which elucidates the choice by using particular redistribution criteria to decompose the prediction into relevance scores. Unlike most other approaches, LRP does not use gradient assessment. The saliency maps generated by GradCAM show the relative importance of each pixels in a given picture [39].

Causal semantics makes sense to explain DNN predictions as explanations must be causal models.

Recent explainable models tend to be unimodal, providing either a visual or a textual explanation. There have also been the first attempts to deliver multimodal explanations. As far as visual question responding and activity identification are concerned, Park et al. [40] were the first to give text and picture explanations. Their methodology, pointing and reasoning explanation (PJE), was devised because of a shortage of acceptable datasets that included human justification.

4.9 Blockchain-based Framework for Trustworthy AI

Despite the fact that XAI seems to be a promising technology, discussing complicated models might lead to a quandary. When an explanation is provided, it automatically leads to the development of a simpler model. Deep learning pioneer Geoffry Hinton warned that explaining AI's workings would be bad. Humans, for the most part, have no notion how they perform and execute their task, and computers are no different. The majority of individuals feel compelled to provide a tale if they are asked to justify their actions. Similar to neural networks, it learns a billion values from training data and uses them to predict things like whether a pedestrian is in a photo. This issue would have been addressed long ago if there were any clear guidelines for determining whether or not a picture included a pedestrian. If many systems are involved in making a prediction and providing consistent explanations, the system is more likely to be trusted.

Polling numerous nodes or predictors that run AI computational models and explain AI findings may help build each predictor's reputation. In this manner, systems may directly or indirectly explain their predictions. A sound rationale for a forecast would add credibility to the final judgment. There must be a mechanism to audit explanation systems in a manner that is irrevocable, unchangeable, and decentralized, as well as one that can be traced and monitored in a way that is very reliable and resilient [41]. Resilience is the ability to withstand manipulation. Changing a block's information breaks its hash, which ties it to the previous block in the chain. Even when all earlier blocks are tampered with, the hash chain is broken since it does not go to the genesis block. Having a full ledger on every node assures that the blockchain will continue to work even if a node fails. Blockchain platforms and decentralized storage systems may help with and deliver these latter functionalities.

4.10 Conclusion

This chapter focuses on explainable AI (XAI). XAI aims to make AI technology more understandable and dependable over time. Explainable artificial intelligence has a long history, although attention has grown in recent years. Much of the gained information may be utilized to comprehend current AI applications, although the models and systems are complicated. This chapter shows different XAI implementations. The next step was to talk about and give several forms of explanations and interpretations. We also stressed the fact that XAI systems that are more secure, trustworthy, and resistant to bias and adversarial assaults may be built on a blockchain-based architecture.

References

[1] F. Mughal, A. Wahid, and M. A. K. Khattak, "Artificial Intelligence: Evolution, Benefits, and Challenges," pp. 59–69, 2022, doi: 10.1007/978-3-030-92054-8_4.

[2] F. Cuzzolin, A. Morelli, B. Cîrstea, and B. J. Sahakian, "Knowing me, knowing you: theory of mind in AI," *Psychological Medicine*, vol. 50, no. 7, pp. 1057–1061, May 2020, doi: 10.1017/S0033291720000835.

[3] N. M. Aliman and L. Kester, "Hybrid Strategies Towards Safe 'Self-Aware' Superintelligent Systems," *Lecture Notes in Computer Science (including subseries Lecture Notes in Artificial Intelligence and Lecture Notes in Bioinformatics)*, vol. 10999 LNAI, pp. 1–11, 2018, doi: 10.1007/978-3-319-97676-1_1.

[4] Q. v. Le, "Building high-level features using large scale unsupervised learning," *ICASSP, IEEE International Conference on Acoustics, Speech and Signal Processing - Proceedings*, pp. 8595–8598, Oct. 2013, doi: 10.1109/ICASSP.2013.6639343.

[5] R. G. Tiwari, A. K. Agarwal, R. K. Kaushal, and N. Kumar, "Prophetic Analysis of Bitcoin price using Machine Learning Approaches," in *Proceedings of IEEE International Conference on Signal Processing,Computing and Control*, 2021, vol. 2021-Octob. doi: 10.1109/ISPCC53510.2021.9609419.

[6] V. Khullar, R. G. Tiwari, A. K. Agarwal, and S. Dutta, *Physiological Signals Based Anxiety Detection Using Ensemble Machine Learning*, vol. 291. 2022. doi: 10.1007/978-981-16-4284-5_53.

[7] M. Westberg, A. Zelvelder, and A. Najjar, "A Historical Perspective on Cognitive Science and Its Influence on XAI Research," *Lecture Notes in Computer Science (including subseries Lecture Notes in Artificial Intelligence and Lecture Notes in Bioinformatics)*, vol. 11763 LNAI, pp. 205–219, 2019, doi: 10.1007/978-3-030-30391-4_12.

[8] F. K. Dosilovic, M. Brcic, and N. Hlupic, "Explainable artificial intelligence: A survey," *2018 41st International Convention on Information and Communication Technology, Electronics and Microelectronics, MIPRO 2018 - Proceedings*, pp. 210–215, Jun. 2018, doi: 10.23919/MIPRO.2018.8400040.

[9] D. Gunning, M. Stefik, J. Choi, T. Miller, S. Stumpf, and G. Z. Yang, "XAI—Explainable artificial intelligence," *Science Robotics*, vol. 4, no. 37, Dec. 2019, doi: 10.1126/SCIROBOTICS.AAY7120.

[10] F. Doshi-Velez and B. Kim, "Towards A Rigorous Science of Interpretable Machine Learning," Feb. 2017, doi: 10.48550/arxiv.1702.08608.

[11] B. Kim, R. Khanna, and O. Koyejo, "Examples are not Enough, Learn to Criticize! Criticism for Interpretability".

[12] W. Samek, T. Wiegand, and K.-R. Müller, "Explainable Artificial Intelligence: Understanding, Visualizing and Interpreting Deep Learning Models," Aug. 2017, doi: 10.48550/arxiv.1708.08296.

[13] B. W. Israelsen, "'I can assure you [$\ldots$] that it's going to be all right' – A definition, case for, and survey of algorithmic assurances in human-autonomy trust relationships," Aug. 2017, doi: 10.48550/arxiv.1708.00495.

[14] M. T. Ribeiro, S. Singh, and C. Guestrin, "'Why should i trust you?' Explaining the predictions of any classifier," *Proceedings of the ACM SIGKDD International Conference on Knowledge Discovery and Data Mining*, vol. 13-17-August-2016, pp. 1135–1144, Aug. 2016, doi: 10.1145/2939672.2939778.

[15] T. Miller, "Explanation in artificial intelligence: Insights from the social sciences," *Artificial Intelligence*, vol. 267, pp. 1–38, Feb. 2019, doi: 10.1016/J.ARTINT.2018.07.007.

[16] A. van Assche and H. Blocked, "Seeing the Forest Through the Trees: Learning a Comprehensible Model from an Ensemble," *Lecture Notes in Computer Science (including subseries Lecture Notes in Artificial Intelligence and Lecture Notes in Bioinformatics)*, vol. 4701 LNAI, pp. 418–429, 2007, doi: 10.1007/978-3-540-74958-5_39.

[17] D. Martens and F. Provost, "Explaining Data-Driven Document Classifications," *MIS Quarterly*, vol. 38, no. 1, pp. 73–100, 2014, [Online]. Available: https://www.jstor.org/stable/26554869

[18] Z. Zhou, Y. Jiang, and S. Chen, "Extracting symbolic rules from trained neural network ensembles," *AI Communications*, vol. 16, pp. 3–15, 2003.

[19] M. D. Zeiler and R. Fergus, "Visualizing and Understanding Convolutional Networks arXiv:1311.2901v3 [cs.CV] 28 Nov 2013," *Computer Vision–ECCV 2014*, vol. 8689, no. PART 1, pp. 818–833, 2014, doi: 10.1007/978-3-319-10590-1_53.

[20] A. Karpathy, J. Johnson, and L. Fei-Fei, "Visualizing and Understanding Recurrent Networks," Jun. 2015, doi: 10.48550/arxiv.1506.02078.

[21] P. Cortez and M. J. Embrechts, "Using sensitivity analysis and visualization techniques to open black box data mining models," *Information Sciences*, vol. 225, pp. 1–17, Mar. 2013, doi: 10.1016/J.INS.2012.10.039.

[22] I. P. Derks and A. de Waal, "A Taxonomy of Explainable Bayesian Networks," *Communications in Computer and Information Science*, vol. 1342, pp. 220–235, 2020, doi: 10.1007/978-3-030-66151-9_14.

[23] R. G. Tiwari, M. Husain, B. Gupta, and A. Agrawal, "Amalgamating contextual information into recommender system," 2010. doi: 10.1109/ICETET.2010.110.

[24] R. G. Tiwari, M. Husain, V. Srivastava, and A. Agrawal, "Web personalization by assimilating usage data and semantics expressed in ontology terms," 2011. doi: 10.1145/1980022.1980133.

[25] R. G. Tiwari, V. Khullar, and K. D. Garg, *Web Recommendation by Exploiting User Profile and Collaborative Filtering*, vol. 832. 2022. doi: 10.1007/978-981-16-8248-3_27.

[26] W. Ma, Z. Huang, C. H. Hsu, and C. E. Montenegro-Marin, "Constraint Programming with Intelligent Backtracking using Artificial Intelligence," *Arabian Journal for Science and Engineering 2022*, pp. 1–11, Jan. 2022, doi: 10.1007/S13369-021-05964-2.

[27] L. Miranda, J. Viterbo, and F. Bernardini, "A survey on the use of machine learning methods in context-aware middlewares for human activity recognition," *Artificial Intelligence Review 2021 55:4*, vol. 55, no. 4, pp. 3369–3400, Oct. 2021, doi: 10.1007/S10462-021-10094-0.

[28] A. Gattami, Q. Bai, and V. Aggarwal, " Reinforcement Learning for Constrained Markov Decision Processes ," in *Proceedings of The 24th International Conference on Artificial Intelligence and Statistics*, Jun. 2021, vol. 130, pp. 2656–2664. [Online]. Available: https://proceedings.mlr.press/v130/gattami21a.html

[29] W. Bannour, A. Maalel, and H. H. ben Ghezala, "Emergency Management Case-Based Reasoning Systems: A Survey of Recent Developments," *Journal of Experimental & Theoretical Artificial Intelligence*, pp. 1–24, Jul. 2021, doi: 10.1080/0952813X.2021.1952654.

[30] X. Wang *et al.*, "Ordering-Based Causal Discovery with Reinforcement Learning," *IJCAI International Joint Conference on Artificial Intelligence*, pp. 3566–3573, May 2021, doi: 10.48550/arxiv.2105.06631.

[31] N. Jussien and V. Barichard, "The PaLM system: explanation-based constraint programming," in *Proceedings of TRICS: Techniques foR Implementing Constraint programming Systems, a post-conference workshop of CP*, 2000, vol. 2000, pp. 118–133.

[32] J. Amilhastre, H. Fargier, and P. Marquis, "Consistency restoration and explanations in dynamic CSPs—Application to configuration," *Artificial Intelligence*, vol. 135, no. 1–2, pp. 199–234, Feb. 2002, doi: 10.1016/S0004-3702(01)00162-X.

[33] B. Y. Lim and A. K. Dey, "Toolkit to Support Intelligibility in Context-Aware Applications," *Proceedings of the 12th ACM international conference on Ubiquitous computing*, 2010, doi: 10.1145/1864349.

[34] F. Elizalde, L. E. Sucar, A. Reyes, and P. Debuen, "An MDP Approach for Explanation Generation.," in *ExaCt*, 2007, pp. 28–33.

[35] R. G. Tiwari, M. Husain, V. Srivastava, and K. Singh, "A hypercube novelty model for comparing E-commerce and M-commerce," 2011. doi: 10.1145/1947940.1948068.

[36] H. Agarwal, P. Tiwari, and R. G. Tiwari, "Exploiting Sensor Fusion for Mobile Robot Localization," 2019. doi: 10.1109/I-SMAC47947.2019.9032653.

[37] R. G. Tiwari, A. Misra, V. Khullar, A. K. Agarwal, S. Gupta, and A. P. Srivastava, "Identifying Microscopic Augmented Images using Pre-Trained Deep Convolutional Neural Networks," 2021. doi: 10.1109/ICTAI53825.2021.9673472.

[38] T. Peltola, "Local Interpretable Model-agnostic Explanations of Bayesian Predictive Models via Kullback-Leibler Projections," Oct. 2018, doi: 10.48550/arxiv.1810.02678.

[39] M. Harradon, J. Druce, and B. Ruttenberg, "Causal Learning and Explanation of Deep Neural Networks via Autoencoded Activations," Feb. 2018, doi: 10.48550/arxiv.1802.00541.

[40] D. H. Park *et al.*, "Multimodal explanations: Justifying decisions and pointing to the evidence," in *Proceedings of the IEEE Conference on Computer Vision and Pattern Recognition*, 2018, pp. 8779–8788.

[41] A. K. Agarwal, R. G. Tiwari, R. K. Kaushal, and N. Kumar, "A Systematic Analysis of Applications Of Blockchain in Healthcare," in *Proceedings of IEEE International Conference on Signal Processing,Computing and Control*, 2021, vol. 2021-Octob. doi:10.1109/ISPC C53510.2021.9609339.

5

Human–Computer Interaction (HCI): An Advancement in the Development of Computer Technology

Shikhar Raghuvanshi[1] and Amit Kumar Tyagi[1,2]

[1]School of Computer Science and Engineering, Vellore Institute of Technology, India
[2]National Institute of Fashion Technology, New Delhi, 110016, India
Email: shikhar.raghuvanshi2020@vitstudent.ac.in;
amitkrtyagi025@gmail.com

Abstract

Human–computer interaction (HCI) mainly focuses on the techniques through which humans connect with various technological developments, data, and other automated chores. This chapter throws light on the significance of HCI, its features, future scope, and possible areas of research. Considering the enhancement and advancements in HCI, it is one of the strongest subjects with high possibilities of research as it is believed to be an area of considerable evolution with human-centric tech strategies that would make our lives easier. Most of the research experiments with regard to HCI are related to the youngsters who are well-educated and technically sound. This research work highlights the mental model of HCI along with various approaches to the same and the trends and patterns in this field. Apart from this, the chapter signifies the aspect of emotional intelligence of the users when it comes to applications that span across the field of HCI. The development and design of an automated system that performs such a task is still being accomplished. This chapter emphasizes human–computer interaction that is a significant aspect of the computer information systems in detail.

Keywords: Human–Computer Interaction, Artificial Intelligence, Internet of Things, Security, Computer Technology.

5.1 Introduction

Human–computer interaction is a vital aspect of computer information systems study that has evolved over the years. It is a study that concerns itself with how humans interact with the computer system. Thus, human–computer interaction is a two-way study involving humans and the computer system [1]. Moreover, the study of the interaction between humans and the computer has stood the test of time, and it is as old as computer technology itself. It is a field that requires the expertise of computer scientists and psychologists. Computer scientists are often concerned with the interaction between computers and humans in which computer comes first before humans, whereas psychologists put humans before computers in their study. The enhancement in technology, developments, and innovative strategies has led to the possibility of bridging the gap between physical humans and virtual interfaces [2]. A user interface is a term used to describe the systems that permit users and humans to interact and collaborate with computers or technical machines. Hardware components that are physical in nature along with the software programs is what constitutes the user interface system as a whole. In fact, it is the user interface that facilitates the channel or medium through which information can be input into the system or output to the user [3].

The purpose of the user interface is to foster an interaction between the user and the machine that will result in effective operations and carrying out of computerized activities. The interaction also ensures that there is feedback and communication in what the user wants to be performed by the machine, which will also aid in decision-making. User interfaces are also designed to make the use of the machine to be an enjoyable and efficient experience [4]. With the continued improvements on technology taking place every day, the user interface has faced modifications that have seen the system become more graphical and user friendly in nature.

The personalization of computer systems with particular focus on the user interfaces has been viewed to provide each individual user with the particular set of information that they need. The personalization of web services has been seen to improve the interaction that the user will have with the information presented by the machine [5]. The growing use of the Internet as a means of sharing and distributing information has initiated the growth of instruction-oriented websites or web pages that offer instructions in the form of supplemental materials. The Internet incorporates the use of hypertext or hypermedia programs in the form of link nodes that provide accessibility to various information sites.

The information contained in the hypertext is usually in a non-linear and an unstructured form, with the main method of data manipulation being scrolling and clicking. The amount of data contained in a web page makes it difficult for the user to view all the information in one sitting when compared to using a textbook [6]. This creates a loss of interaction effectiveness between the user and the computer; such a challenge presents a problem to the designers of software programs who find it difficult bridging the gap of effectiveness in the user interface. Hypertext functionality provides approaches for finding information over the Internet. The most common approach is typing the URL link of the information that one is interested in within the browser location after which the page shows the information requested by the user [7]. The user mostly has to describe the information they are looking for by keying in a search word or letter. This however presents a problem when the results from the web page display information that the user was not looking for or the requested information is too short and unclear. Another problem presents itself when the webpage provides interesting information, making it difficult for the user to navigate through the page and find the most useful piece of data [8].

The human–computer interaction mechanism is a two-way framework that involves both humans and computerized systems. This interaction dates back to the days of ENIAC in the 1940s, which involved programmers, operators, managers, etc. The very origin of this concept revolves around the core topic of information gathering and analysis and the human elements such as imagery, statistical analysis, motion pictures, etc. [9]. The philosophical and scientific aspect put forth in 1914 has truly set up the pace for HCI. The human–computer interaction has attained drastic improvement in computing. The advent of graphical user interface brought a revolution that transformed human–computer interaction. Programming is another significant aspect of human–computer interaction communication between the programmer and the computer system [10]. Desktop computers that were considered great achievement have been shifted to portable devices such as smartphones. Consequently, this has made the communication easier and seamless. Hardware and software technologies are associated with human beings and computer systems. The key pillars of current human–computer interaction are linked with computer operation, information systems management, and discretionary use. However, development has not reached its apex and has the potential of further advancement that will increase the relationship between humans and computer system in future [11].

5.2 History of Human-Computer Interaction

The human-computer interaction is focused on three basic fields that include computer operation, information systems management, and discretionary use. Human-computer interaction became a known field of research in relatively few decades ago; however, people have related to computer systems for a very long time [12]. The history of human-computer interaction is the same as the history of computer system and its usage by humans that linked to the early days of the ENIAC in the 1940s. Here, the human-computer interaction can be broadly classified into three major roles that include managers, programmers, and operators. The early advancement in human-computer interaction is discussed in a topic entitled "man-computer symbiosis." He further stated that he hoped that in not many years thence, the human brain will be closely coupled with computing machines. The resultant of the partnership will be a system that will be able to think beyond the human brain and also process data in a specific way that is far better than the computer systems of yesteryears [13].

The origin of human–computer interaction also involves the origins of human factors and the origins of the focus on information. The human factor aspect of HCI originated in the late 19th century from technologies such as photography, statistical analysis, and moving pictures. These improvements came as a result of the need to improve work practice by reducing performance time. The work of Lillian Gilbreth and her husband in 1914 set the pace in the combination of scientific management and philosophy that birthed human-computer interaction. Early technology and philosophy truly set the pace for human–computer interaction as we see it today [14].

5.2.1 Present day of human-computer interaction

Of course, human–computer interaction is an ever-evolving field. The advent of graphical user interface brought a revolution in the computing world. The graphical user interface was a disruptive revolution that transformed human–computer interaction [15]. However, as time progressed, the graphical user interface becomes more acceptable, and revolution became less disruptive. The revolution transformed every aspect of a computer system, from the use of computer systems at home to their use in organizations. In addition, the move from organizational focus to the web and broader end-user computing is also a major move in human–computer interaction.

The human–computer interface is built into a task situation and environment. The task can vary depending on what the user intends to do with the computer system. The overlapped area represents the aspect of the interaction involving the activities of the human and the processes of the computer [16]. The non-overlapped area in the human part involved cognitive processes involved in the task. But it does not require computing processes of the machine, while the non-overlapping process in the computer represents computer processes that are not related directly to the interface. In human–computer interaction, information flows from the user to the computer and also from the computer to the user. Human–computer interaction is improving by leaps and bounds especially with the current improvement in software and hardware technologies. The health of technology industry is closely related to the current innovation in hardware technologies. Moreover, the boost in these technologies as well as in human–computer interaction can be interpreted from the Moore's law that can be applied in a broader scope to represent the exponential growth in this field. Human–computer interaction today still reflects the three major roles of HCI obvious in the early days of computing. These roles including system analysis and management as well as computer operation and programming are now the focus of human–computer interaction research works. Even though a lot of evolutions have taken place in these roles, some aspects of these early roles are still obvious in human–computer interaction [17].

Programming is a vital aspect of human–computer interaction as it provides a platform for effective communication between the programmer and the computer system. Interest in programming as a human activity increased as a result of the software crisis. Furthermore, the development of interactive computing and time sharing made it possible for new styles of programming to evolve. As time evolved, programming and system development became more centered on the user. A lot of advances have been seen in computing today and of course in human–computer interaction [18]. Desktop computers that were the heights of the technology a few years ago have lost their spotlights to portable devices such as smartphones. This has made the communication and hence the interaction between human and computers much easier and seamless. The future still holds much more prospects to advancement in these technologies and the relationship between humans and computer system.

5.2.2 Applications of HCI in different real-life sectors

Today, technology has become a part of every part of our lives, and we use it all the time. It does not even matter if someone does not own or use a computer. Computers still have an impact on their lives even if they do not know it. Some examples of computer interfaces that people can use every day without having a computer are ATM machines, train ticket machines, and hot drink vending machines. HCI is very important when you make any of these systems or interfaces. To make sure the interface is safe, useful, and efficient, users need to think about HCI principles [19–21].

5.3 Industry and Business

They need to pay attention to how they use technology or computers in their jobs every day. A well-designed, simple system makes staff happier and more productive because it does not make them angry, which makes them happier and more productive. There are a lot of safety-critical systems, like power plants and air traffic control rooms, which need to be designed with HCI in mind. The death of many people can happen when there are design flaws.

5.3.1 Accessibility

In order to make systems that are both functional and accessible to people with disabilities, human–computer interaction (HCI) is a good thing to think about when making them. HCI's main goal is to make sure that everyone has safe, usable, and efficient systems that can be used by all people. The people in this group have a wide range of skills and know-how, and they come from all walks of life. If a system is made with user-centered approaches and concepts, people with disabilities will be able to use it as much as they can, too.

5.3.2 Users who are untrained

Only a very small number of computer users read the manual for the software, if there is one. It is not just the most specialized and complicated programmers that need a lot of training and help. People think they should be able to figure out how a computer programmer works in a few minutes. HCI gives designers the ideas, strategies, and resources they need to make interfaces that are easy to use and do not need any training.

5.3.3 Other applications

In the workplace, virtual reality is a great example of how humans and computers can work together. There should be a new way of looking at things because of how people work together with a computer and see things. VR, when done right, can make things look and feel like the real thing, and it can even be fun. This is a good example of human–computer interaction (HCI) at work. Voice search, such as Amazon Alexa and Google Voice Search, can also be used to find things. They speak to a device or system that leads them to an Amazon buy or a Google search.

5.4 Recent Development in Human-Computer Interaction

Looking at what the future of human–computer interaction looks like is a good way to keep up with what is going on. This is because human–computer interaction (HCI) wants technology to be as normal and easy to use as possible in our daily lives, giving us a glimpse into what might seem like the future. Many businesses have used HCI theories to come up with prototypes that are very new.

5.4.1 Dexta haptic gloves

When people use virtual reality, Dexta haptic gloves can make people feel hardness, softness, springiness, and more. People who use VR gloves feel like they are moving their fingers in a way that is similar to how they feel when they interact with things in the virtual world. Find out how VR can be used in the military by reading this.

5.4.2 Pre-touch sensing

They move their fingers in a way that is similar to how they move their fingers when they interact with things in the virtual world when they use VR gloves, so they feel the same way.

5.4.3 Paper ID

It is the next step in the process of converting paper to a touchscreen, and it is done. If this new technology is used, paper will be able to "see where it is and follow gestures," as well as be able to connect to the Internet of Things and other things. There is a goal to connect the real world and the digital world. Soon, HCI will be used in many different ways. Think of the door as a window to the future, and can be pretty sure there is a lot more inside.

5.5 Evaluation Methods and Metrics for HCI

People have used the thesis of someone who is a doctor to help them with this, total value of the UEM has been calculated, taking into account things like its usefulness, need for training, and how much effort it takes to use it and how much it helps [22–24]. The strengths and flaws of different methods were also taken into account when choosing the UEM. Even other studies were taken into account, like people's own experiences and what they said in those studies. There were a lot of UEMs that were chosen for inspection and testing.

Heuristic evaluation: Useful engineering is a way to look for usability problems in a user interface design so that they can be fixed as part of a design process that will keep going back and forth over and over again. It is done by having a small group of people look at the interface and say how well it meets known usability rules (the "heuristics")

Cognitive walkthrough: Usability testers do a set of tasks and ask questions from the point of view of a person to see how it works. Researchers want to find out how easy it is for new people to learn the system through a "cognitive walk through."

Formal experiments: A controlled experiment with people who will be used as test subjects is what these tests are all about. There were people who were watching to see how the users did what they were told. A lot of information that can be used for later analysis can be found in this text (video with the actions and reactions of the user). If users look at how long it took and how many mistakes were made when users did something, users can do a statistical analysis of what they did.

Constructive interaction: During this method, two people work together to learn about the system they are working on at the same time they talk to each other about how they feel, like having a conversation. It is not just for a set of tasks that users talk and interact with each other about.

Interviews: People who use the interactive system and people who evaluate it talk to each other in this way all the time. People do not look at this, but the person who interviews them does. People who use the system are interviewed to get information about how they feel about it, what they think about it, and so on.

Questionnaires: During an evaluation, the person in charge asks people how the system works so that they can figure out what it is like. In order to get both qualitative and quantitative information about how the users feel, the questionnaires ask for both qualitative and quantitative information.

5.6 Ethical, Legal, and Social Issues of HCI

People who are terrible can take advantage of the law and ethics of computers in order to accomplish things that are not beneficial for the world [25]. Many individuals are concerned about the unlawful downloads of music and videos, as well as the presence of spyware on their computers. On top of that, adolescents are also dealing with issues such as identity theft and cyberbullying, which they must also contend with.

It is feasible to have computers that collaborate with other individuals. In the second part of the 1990s, there was a significant increase in the amount of time people spent using computers in groups. Listservs, newsgroups, bulletin boards, and chat rooms, all of which are available online, allow the groups to connect with one another and share resources with one another. A large number of research works have demonstrated that this is not the case.

In the field of human–computer interaction, ethics covers a wide range of topics, including human wellbeing, property rights, privacy, freedom from bias, universal usability, trust, autonomy, and informed consent. Many researchers and practitioners in biomedical informatics are worried about the ethical implications of technology design and use, much as researchers and practitioners in human–computer interaction are concerned about the ethical implications of technology design and usage. As previously stated, the breadth of ethical considerations in various domains can be considered extensive. Certain topics are more prevalent in literature than others at any particular moment, and this is true for all genres.

Numerous individuals within biomedical informatics are interested in investigating the ethical aspects of human–computer interaction, which is a growing field. In recent years, biomedical informatics research has focused on the ethical difficulties that occur when patients utilize tools such as mobile devices and social media to perform activities such as seek medical care and connect with their doctor [26]. These topics include privacy, inclusion, and e-consent. Because of the advent of personalized medicine and citizen science, there will be an increase in the number of questions that people want to know the answers to.

5.7 Future of Human-Computer Interaction

It is a well-known fact that technology is maturing faster than we ever imagine. In several work, the authors mentioned that governments and various organizations are now investing a lot of money on high-performance computing, and we can expect an exceptional advancement in this field. Also, few researchers outlined that human–computer interaction can be seen in various fields that include direct manipulation of graphical objects, the mouse, windows, and so forth. These provide a medium for effective interaction with the computer [27]. This is essentially used in different kinds of applications today. Some of these applications include drawing programs, text editing programs such as word processors like MS Word, and so forth. It also includes spreadsheets such as MS Excel, HyperText, computer-aided design (CAD) such as Autocad, video games, and so forth. However, much more effective interaction is expected in the future as this field evolves. These include gesture recognition, more effective multi-media platforms, three-dimensional platforms, computer supported cooperative work, virtual reality and augmented reality, natural language and speech, and so forth.

As pointed out by scientists, human–computer interaction will continue to evolve. Since the 1960s, the core areas of HCI study include software engineering, software human factors, computer graphics, and cognitive science. These fields have greatly evolved and will continue to evolve. Moreover, these important fields of human–computer interactions have essentially intertwined with each other over the years. More so, a number of other interesting fields have been incorporated into human–computer interaction, and these include networking, broader aspects of social science, media, information management, and artificial intelligence [28]. But this does not stop here as the future holds many prospects for human–computer interaction. In the words of Carroll (2011), the future promises to be even more exciting than the past in the field of human–computer interaction. It is projected greatly to influence all other aspects of computer science and also technology in general, given the relevance of computing to technology today and the world in general.

5.8 Methodology and Scope of the Research

There are a number of methodologies that have been used in human–computer interactions and in developing interaction design. These methodologies have their foundations from early designs that viewed human cognitive processes as predictable and quantifiable. These characteristics saw computer engineers and interface developers focusing on cognitive psychology and behavioral science when designing user interfaces. The current models have focused on constant feedback that exists between computer users and designers [29]. The design methodologies that are used in HCI include design research, research analysis and concept generation, prototyping and usability testing, and system testing. Design research involves using techniques such as interviews, observations, or questionnaires to gather the existing information on HCI. The designers of user interfaces investigate users so as to learn more about their interactions with the computers and how this interaction can be improved. For the purposes of this report, the design research that will be used will be secondary data collected from academic books, publications, journals, and web articles that contain information on human–computer interaction and web-based systems.

Research analysis and concept generation focuses on the design research collected from the observations, interviews, or questionnaires. The design research is usually used to create concepts that will be used in software and hardware development. Research analysis of the HCI process involves the refinement of the design research to suit the design needs of the interface. Research design analysis is usually conducted by brainstorming and discussion to come up with appropriate research information that will be used in developing the user interface. Once the design research has been analyzed, the next stage will involve concept generation where the most suitable analysis is chosen for developing the user interface. Prototyping and usability design methodologies are used to test the aspect of the concepts that have been generated. Prototypes are used to determine whether the user interfaces have met the requirements of the human users as well as the collected design information. Prototype testing is also performed to determine whether the user interface will be usable. System testing is performed once the system has been completed and it is used for identifying any errors that might occur in the interface [30]. The scope of research in this report has focused on the human–computer interactions in web-based systems. The theory and fields of study that underlie HCI have been examined as well as the types of HCI that are in use for web-based systems. HCI studies and research work has been conducted to create a better understanding of the cognitive behavior of human users and the impact of this behavior on the computer's operations. The research findings and discussion have shown that HCI is mostly concerned with the individual user as well as a larger context of computer users. Research findings have also shown that HCI research work has mostly focused on improving the existing versions of user interfaces to a more virtual reality status.

5.9 Recommendations for Future

To improve existing human–computer interactions, continuous research work needs to be conducted to develop newer and better user interfaces that are in line with current technological innovations. User modeling concepts should also be used to improve the human–machine interaction by offering solutions to the problem of searching for relevant information [31]. User modeling involves gathering information about the user from the search information they have keyed in that will be used to change the content of the webpage and the links that display results of the user's request. User modeling is meant to capture the thinking process and knowledge of the machine user that will be integrated into the URL links so that the users search request can be determined and processed. Adaptive hypermedia can also be used to improve the human–computer interaction by providing web personalization services for every user of the webpage. Personalization has taken the form of registering user information with particular websites by providing information about the user's interests and opinions. Once the user registers, they can personalize the webpage and the information contained in the website to suit their personal needs. The website in turn monitors the user's patterns and behavior that are used to develop a user model that represents the user's interests [32–36].

5.10 Conclusion

The field of human–computer interaction is no doubt a complex one but vital in bridging the gap between the user and the computer. There is a growing emphasis on designing interfaces that will meet the needs, program abilities, and skills of the user as well as provide a personalized experience when using web-based programs and applications. There is a need for designers and architects of web-based applications to view interactive systems as important in ensuring the success of the program. This can be accomplished by ensuring there is compatibility between the user interface design and the skills, knowledge, and characteristics of the user by analyzing their cognitive processes; this will ensure that the human–computer interaction is effective and efficient.

References

[1] anogaran, Gunasekaran, Chandu Thota, and Daphne Lopez. "Human-computer interaction with big data analytics." *Research Anthology on Big Data Analytics, Architectures, and Applications*. IGI global, 2022. 1578-1596.

[2] iele, Cezary. *Human Movements in Human-computer Interaction (HCI)*. Springer, 2022.

[3] iu, Rui, et al. "Multi-Stage Deep Transfer Learning for EmIoT-enabled Human-Computer Interaction." *IEEE Internet of Things Journal* (2022).

[4] remalatha, R., and R. S. Aiswarya. "Physical education system and training framework based on human–computer interaction for augmentative and alternative communication." *International Journal of Speech Technology* (2022): 1-11.

[5] arahasanović, Amela, and Alma Leora Culén. "Project-based learning in human–computer interaction: a service-dominant logic approach." *Interactive Technology and Smart Education* (2022).

[6] un, Ying, et al. "Context-Aware Augmented Reality Using Human–Computer Interaction Models." *Journal of Control and Decision* (2022): 1-14.

[7] iederich, Stephan, et al. "On the design of and interaction with conversational agents: An organizing and assessing review of human-computer interaction research." *Journal of the Association for Information Systems* 23.1 (2022): 96-138.

[8] adley, Greg, et al. "The Future of Emotion in Human-Computer Interaction." (2022).

[9] 'Brien, Heather L., et al. "Rethinking (Dis) engagement in human-computer interaction." *Computers in Human Behavior* 128 (2022): 107109.

[10] Subba, Tanuja, and Tejbanta Singh Chingtham. "A Survey on Biosignals as a Means of Human Computer Interaction." *Contemporary Issues in Communication, Cloud and Big Data Analytics*. Springer, Singapore, 2022. 297-306.

[11] Liu, Xin, and Hua Pan. "The Path of Film and Television Animation Creation Using Virtual Reality Technology under the Artificial Intelligence." *Scientific Programming* 2022 (2022).

[12] Subba, Tanuja, and Tejbanta Singh Chingtham. "A Review on Types of Machine Learning Techniques for Biosignal Evaluation for Human Computer Interaction." *Advanced Computational Paradigms and Hybrid Intelligent Computing*. Springer, Singapore, 2022. 457-466.

[13] Fairclough, Stephen H. "Designing human-computer interaction with neuroadaptive technology." Current Research in Neuroadaptive Technology. Academic Press, 2022. 1-15.

[14] Costa, Simone Dornelas, et al. "A core ontology on the Human-Computer Interaction phenomenon." *Data & Knowledge Engineering* (2022): 101977.

[15] Baumann, Konrad, et al. "HCI-E $$∧ 2$$: HCI Engineering Education." (2022).

[16] Carmichael, Dawn, Claire MacEachen, and Jacqueline Archibald. "Gamification in a Learning Resource for the Study of Human Computer Interaction." *Intelligent Computing*. Springer, Cham, 2022. 697-717.

[17] Giannakos, Michail, et al. "Lots done, more to do': The current state of interaction design and children research and future directions." *International Journal of Child-Computer Interaction* (2022): 100469.

[18] Chua, S. N., et al. "Hand Gesture Control for Human–Computer Interaction with Deep Learning." *Journal of Electrical Engineering & Technology* (2022): 1-10.

[19] Kaptelinin, Victor. "The social production of technological autonomy." *Human–Computer Interaction* (2022): 1-3.

[20] Krol, Laurens R., and Thorsten O. Zander. "Defining neuroadaptive technology: the trouble with implicit human-computer interaction." *Current Research in Neuroadaptive Technology*. Academic Press, 2022. 17-42.

[21] Kandemir, Hasan, and Hatice Kose. "Development of adaptive human–computer interaction games to evaluate attention." *Robotica* 40.1 (2022): 56-76.

[22] Mattyasovszky-Philipp, Dóra, Attila Márton Putnoki, and Bálint Molnár. "The Unrepeatable Human Mind—Challenges in the Development of Cognitive Information Systems—What Makes a Machine Human?" *Electronics* 11.3 (2022): 394.

[23] Šumak, Boštjan, Saša Brdnik, and Maja Pušnik. "Sensors and Artificial Intelligence Methods and Algorithms for Human–Computer Intelligent Interaction: A Systematic Mapping Study." *Sensors* 22.1 (2022): 20.

[24] Naim, Mohd, and Nur Faraha. "The Pragmatic Idea of HCI Notion and KE Approach in Video Mobile Learning." *Proceedings of Sixth International Congress on Information and Communication Technology*. Springer, Singapore, 2022.

[25] Biele, Cezary. "Hand Movements Using Keyboard and Mouse." *Human Movements in Human-Computer Interaction (HCI)*. Springer, Cham, 2022. 39-51.

[26] Kaveri, V. Vijeya, et al. "Object Tracking Glove." *Materials Today: Proceedings* (2022).

[27] Biele, Cezary. "Movement in the Environment." *Human Movements in Human-Computer Interaction (HCI)*. Springer, Cham, 2022. 89-102.

[28] Siovi, Kelvin Wambani. A Framework for Design Thinking of Gesture-Based Human Computer Interactions. Diss. JKUAT-COETEC, 2022.

[29] Reig, Samantha, et al. "Theory and Design Considerations for the User Experience of Smart Environments." *IEEE Transactions on Human-Machine Systems* (2022).

[30] Biele, Cezary. "Perception of Movement." *Human Movements in Human-Computer Interaction (HCI)*. Springer, Cham, 2022. 103-115.

[31] Shibuya, Yuya, Andrea Hamm, and Teresa Cerratto Pargman. "Mapping HCI research methods for studying social media interaction: A systematic literature review." *Computers in Human Behavior* 129 (2022): 107131.

[32] Biele, Cezary. "Leg and Foot Movement." *Human Movements in Human-Computer Interaction (HCI)*. Springer, Cham, 2022. 53-62.

[33] M. M. Nair, A. K. Tyagi and N. Sreenath, "The Future with Industry 4.0 at the Core of Society 5.0: Open Issues, Future Opportunities and Challenges," 2021 International Conference on Computer Communication and Informatics (ICCCI), 2021, pp. 1-7, doi:10.1109/ICCCI50826.2021.9402498.

[34] Tyagi A.K., Fernandez T.F., Mishra S., Kumari S. (2021) Intelligent Automation Systems at the Core of Industry 4.0. In: Abraham A., Piuri V., Gandhi N., Siarry P., Kaklauskas A., Madureira A. (eds) Intelligent Systems Design and Applications. ISDA 2020. Advances in Intelligent Systems and Computing, vol 1351. Springer, Cham. https://doi.org/10.1007/978-3-030-71187-0_1

[35] Goyal, Deepti & Tyagi, Amit. (2020). A Look at Top 35 Problems in the Computer Science Field for the Next Decade. 10.1201/9781003052098-40.

[36] Madhav, A.V.S., Tyagi, A.K. (2023). Explainable Artificial Intelligence (XAI): Connecting Artificial Decision-Making and Human Trust in Autonomous Vehicles. In: Singh, P.K., Wierzchoń, S.T., Tanwar, S., Rodrigues, J.J.P.C., Ganzha, M. (eds) Proceedings of Third International Conference on Computing, Communications, and Cyber-Security. Lecture Notes in Networks and Systems, vol 421. Springer, Singapore. https://doi.org/10.1007/978-981-19-1142-2_10

6

Human–Computer Interaction and Explainable Artificial Intelligence: For the Effective Use of Technology

Arun Viswanathan[1] and Amit Kumar Tyagi[2]

[1]School of Computer Science and Engineering, Vellore Institute of Technology, India
[2]National Institute of Fashion Technology, New Delhi, 110016, India
Email: arun.viswanathan2020@vitstudent.ac.in amitkrtyagi025@gmail.com

Abstract

Human–computer interaction (HCI) is research into how humans and computers interact and communicate with each other. It focuses on how people use computers as well as how computers use people. A group of people who study how humans interact with computers is called HCI. They look at how humans interact with computers and come up with new technologies that let people do so in new ways. In the beginning, HCI focused on computers, but now it is a broad field that deals with many different types of information technology design. As computer technology moves toward making products available to everyone in the world, the need for efficient interaction between humans and computer devices for people who are less skilled grows more important. From the start, HCI would grow to include a lot of different fields, like computer science, cognitive science, and human-factors engineering. In today's world, many people question and debate whether or not humans should physically do simple tasks. Every part of our world is becoming more and more automated and mechanized. An intelligent person must give this automation a sense of direction. We need a well-researched view of human–computer interaction to be able to seamlessly translate our views and instructions into machine input. This chapter talks about the concept of human–computer interaction (HCI) and its branches, like user interface design and user experience design, which are looked at in depth. A needful goal for research and developing human-computer interaction system is to make system human-human intractable and intelligent. Much of this chapter also talks about what is going on with the subject now and how it works in different fields like artificial intelligence, home automation, information technology, and so on. To learn more about how human-computer interaction system are being used in Industry 4.0 and how it is different from other types of interaction systems, such information will be discussed in this chapter.

Keywords: Human-Computer Interaction, User Experience, User Interface Design, Artificial Intelligence, Automation, Industry 4.0.

6.1 Introduction

Human–computer interaction (HCI): This is the research and study of how humans and computers interact with each other. It is a cross-discipline, which covers a wide range of contents. HCI, primarily a computer science field, has also expanded to other fields such as cognitive science, linguistics, behavioral and cultural anthropology, and design. HCI is of most use in the academic sense and as a base for greater technology to build form. It involves itself in research and understanding of how humans use technology and how they are affected by it.

Artificial intelligence: Artificial intelligence is a branch of computer science that is concerned with the simulation of human intelligence as computer programs to perform tasks and aid human life. The application of an artificially intelligent machine is plenty and also well known. The development in fields like medicine, education, transportation (automobile industry), etc. will be exponential. The basic necessity of a good AI machine is seamless and flawless human–computer interface. What good is AI if it does not serve useful to humans?

Explainable artificial intelligence (XAI): XAI is a collection of techniques that allow humans/users to analyze and understand the reliable results obtained as an output of machine learning algorithms. XAI is a term that describes the possible biased nature of a model and its likely impact. When it comes to AI-assisted decision-making processes, it complements and supports the processes of evaluating the correctness, fairness, transparency, and outputs of models. An organization's capability to elucidate AI models is extremely important when it comes to the case of deploying and production of AI models. It also supports and strengthens situations of responsible AI strategies for a certain organization.

For humans, comprehending and retracing the steps followed by evolution of AI along with arriving at a suitable conclusion can be quite tedious. These calculation processes are turned into a complete black box that cannot be deciphered. These black-box models are curated with the help of sufficient data, making it difficult for even the engineers, program developers, and data scientists to decode the internal functioning of the black box. There are various benefits to understanding how an AI-enabled system arrived at a specific conclusion. Explainability can help developers ensure that the system is working as planned, it may be required to meet regulatory requirements, or it may be crucial in allowing parties affected by a decision to protest or change the decision.

Automation: Automation is another sub-branch in computer science related to AI and ML. It is the process of decreasing human input in any tasks be it novel or complex. The advancement into the mechanized world demands the technology that requires lesser and lesser input from humans. With the click of a button or swipe of a finger, we want our menial tasks to be done. Home automation and automated factories and production units are concepts that again require human–computer interaction.

User experience: User experience, when compared to HCI, is a more practical and consumer-based aspect of the discipline. It concerns itself with design and aesthetics and focuses on the experience of the user when interacting with technology. There are many tenets to user interaction and the process is ongoing. UX designers gain insights regularly from users, which is applicable to the continuous development of technology.

Technology has a large reach and a strong grip on the processes that the world runs on today. As these processes are rapidly getting automated, the risk of human input becoming sequestered increases. Human–computer interaction and user experience are predominately concerned with keeping humans at the center of the systems and technologies.

Industry 4.0: Industry 4.0 is the transformation of the way we produce things with the introduction of digitization to manufacturing. New technology like IoT, AI, and machine learning paves the way for Industry 4.0. The use of sensors and embedded software and robotics to collect and analyze data are a key factor in these smart factories. The usage of such features is built upon good HCI.

6.2 Overview of HCI and XAI

With the rise of HCI over the last decade, it is now impossible to tell what is real and what is not. Human–computer interaction (HCI) has three parts: a human, a computer, and their interactions. Humans use computer inferences to help them reach their goals. An interface is the part of a computer that lets anyone talk to it. In HCI, interfaces play a big part, and they help people do their jobs. We need to know what each part can and cannot do in order to build efficient and effective interfaces. Information is put in and taken out in different ways by humans and machines.

Human–computer interaction (HCI) is a branch of computer science that investigates the aspects of interaction between users and computers. It also looks into how computer systems can be made so that they can be used by people who have specific needs and habits. Computer science, behavioral sciences, and design are all part of this field that combines them all. To make computer systems easier to use and understand is one of the main goals of HCI. Users interact with computers through a "user interface," which is made up of hardware and software that lets users change the system and the system tell them what they need to know about the system. When it comes to the design, implementation, and evaluation of user interfaces, HCI places a lot of importance on these things and how they work.

According to HCI, a good theory or model of human–computer interaction is required for good interface design. This theory or model should be based in large part on a theory of human cognition to model the cognitive processes of users when they use computer systems. Cognitive psychology or cognitive science, which is a multi-disciplinary field, is often used to make these kinds of models of human cognition. While philosophers have not studied HCI directly, they have made a lot of contributions to the theory of cognition, especially the relationship between cognition and the outside world, and this is where philosophy and HCI meet.

Explainable AI (XAI), also known as interpretable AI, is artificial intelligence (AI) that allows people to understand the results of the solution. It is different from the "black box" idea in machine learning, where even the creators cannot explain why an AI made a certain decision. XAI could show how the social right to explanation is being used. Users of a product or service can have a better experience even if there is no legal or regulatory requirement for XAI. XAI can help end users believe that the AI is making good decisions.

White-box and black-box machine learning (ML) algorithms are the two types of algorithms used in AI. White-box models are machine learning models that produce results that can be understood by people who work in the field. Black-box models, on the other hand, are very hard to explain and even experts in the field do not understand them. XAI algorithms are said to be transparent, interpretable, and easy to explain or understand. If the approach designer can explain and justify the steps that extract model parameters from training data and make labels from testing data, then the process is transparent. Interpretability is the ability to understand the ML model and explain why it made a decision in a way that a person can understand. Explainability is a term that many people agree is important, but there is no agreed-upon definition. When it comes to machine learning, "explainability" means "the collection of features from the interpretable domain that led to a decision (e.g., classification or regression) for a given example." This means that if algorithms meet these criteria, they can be used to justify decisions and track and so verify them. They can also be used to improve algorithms and look into new facts.

The command "maximize accuracy of assessing how favorable film reviews are in the test dataset" is an example of how AI systems try to act in a way that will help them reach a goal that was chosen by the system's creators. This way, the AI can learn general rules, such as "reviews that use the word 'terrible' are more likely to be bad." But it could also learn bad rules, like "reviews with 'Daniel Day-Lewis' are usually good." These rules might not be good if they do not apply outside of the test set, or if people think the rule is unfair or "cheating." To find out how likely it is that the system will work well with data outside of the test set, a human can look over the rules of an XAI. Since incorrect predictions can cost a lot of money, this is especially important for AI technologies that are used in medicine. XAI has the ability to make algorithms more stable and boost the confidence of medical professionals.

6.3 Methodologies/Tools and Technologies of HCI and XAI Technologies

Not all technology is available to the public with the current economy and the cost of such products playing a huge part. But the majority of these technologies have found their way into day-to-day devices like mobile phones, cars, entertainment systems, etc.

6.4 HCI-based Tools/Technologies

Visual-based HCI:

Visual-based HCI is, by far, the most popular area in research because of its widespread application and variety of solutions it offers to real-world problems. The visual signals in human behavior are recognized and taken as inputs by the machines. Main research areas of visual-based HCI systems right now are:

- Facial expression analysis
- Body movement analysis
- Gesture control
- Eye movement tracking

Even though these systems all use visual-based inputs, they differ in respects to their application. Facial expression analysis is used to recognize human emotions and facial expressions. Gesture control and body movement analysis is the most popular in this area and is used for direct communication/interaction with computers. Eye movement tracking does not have direct applications but is used to work in tandem with other systems to help understand where the user's focus and intent really lies. Eye tracking technology does help people with auditory disabilities.

Audio-based HCI:

Audio-based HCI is another very important area of HCI system. Audio signals in today's world are of an abundant number and using such signals is of paramount importance for an efficient machine. Audio signals are more reliable than visual-based inputs and are more popular in today's world.

Research under the audio-based systems classification is:

- Speech recognition [2]
- Speaker recognition
- Auditory emotion recognition [3]
- Music recognition

Though speech recognition and speaker recognition technology are not perfect [4], it is the most popular form of interaction as this is the easiest for humans to use. To integrate some human emotion in machines, we need the machine to understand what we tell it and also to respond using human emotions. Sounds like gasps, sighs, etc. need to be understood by the machine to work perfectly.

Sensor-based HCI:

This is a system that is of predominant use today. This system, however, requires at least one physical sensor that is used to communicate between the machine and the user. Some of these sensors are:

- Mouse and keyboard
- Joysticks
- Smart pens
- Pressure sensors
- Haptics sensors
- Taste/smell sensors

Several such sensors have been around for a long time and have been used extensively. Haptic and pressure sensors are the new technological advances in this sector. They are used in robotics, virtual reality, video game systems, mobile phones, and automobiles. Sensors are used in the movie and the video game industry for motion capture and computer graphics.

6.5 XAI Techniques

The decision-making process of ML algorithms and AI systems have been questioned extensively and rightly so. The problem is clear and obvious. What biases and what data have driven the algorithm to this conclusion? Interpreting and understanding the data that drives the decision is arguably as important as the decision itself. This predicament has brought a heavy demand of explanations among users. Even though there is no strict definition of interpretation and explanation of algorithms, many methods and models have been developed. This is what drives the whole research field of explainable AI [5, 6].

Usually when creating a data model, a tradeoff can be observed. We find that models that are very interpretable are not very accurate and models that are accurate are not easily interpretable. Simple linear models are easily interpretable by humans but do not give accurate predictions for complex problems and, on the other hand, we build highly complex nonlinear problems that perform better on tasks but are simply too difficult for humans to understand, like neural networks. Based on this, we have two approaches to building AI and ML model based and post-hoc. The former only builds trained models that can be interpreted by humans and the latter focuses on deriving explanations for any complex models. Post-hoc models are further divided into black-box approach and white-box approach. The black-box approach means that we only use the input and output to derive explanations for the model. The white-box approach allows us access to the contents of the model to derive our explanations. The explainable AI field also delves into the psychology of which explanations are better and what makes good explanations as well.

There are a couple of properties that are used to distinguish different types of methods. First is agnosticity. Agnosticity divides models into two groups – model-agnostic and model-specific. Model-specific means that the tools used are specific to only a single model or a group of similar models. Model-agnostic means that the tools used can be used on any ML model no matter the complexity. The next property is scope. Scope categorizes methods into local and global approaches. Sometimes we aim to explain only parts of model (local approach) and sometimes the whole model (global approach). To explain further, the local approach establishes a decision boundary and explains predictions made within it. This is done to make sense of complex nonlinear model where we cannot explain the model as a whole. Then there is data type and explanation type, which are properties that explain what data type the model works on and what type of explanation the model produces like visual explanation, feature importance explanation, data point explanation, or surrogate model explanations. Now that we know some of the terminology, we can look into the different methods used to explain models.

6.6 Feature Importance

Feature importance is a technique that calculates a score for all the input features for a given model. This score calculates the "importance" of a feature. The higher the score is, the more important it is and more will be its effect on the model prediction. Understanding the priorities of the features of a model helps you understand why decisions that have been made were made. Feature importance also helps in the performance of the model. Features of lower importance that do not influence the outcome in a substantial manner are removed, making the model faster. Feature importance makes the model more interpretable.

6.7 Partial Dependence Plot

Partial dependence plot or PDP is a global, model-agnostic explanation method. It shows us how individual features contribute to the target response of the black-box test. To put simply, how does the feature affect the model's decision? It plots a graph where the chosen feature appears on the x-axis and the degree of partial dependence is plotted on the y-axis. This shows us to what extent the outcome depends on the feature.

6.8 Local Interpretable Model-agnostic Explanations (LIME)

As we have discussed earlier, it is important to explain the prediction made by models to humans. However, some problems are too complex to explain as a whole. So, LIME is a technique that uses the local approach and is model-agnostic. The intuition behind LIME is that complex data are slightly perturbed to make it easier for humans to understand. The perturbed data locally is easier to approximate a black-box model than the original data globally. This is done by weighting the perturbed data by their similarity to the original instance that we want to explain.

6.9 SHapley Additive exPlanations (SHAP)

SHAP is also a local model-agnostic method to interpret models. It is based on the Shapley values. The Shapley value is originally from game theory where it measured each player's contribution and has been adopted and represented as an additive feature attribution method, therefore making it a linear method. The SHAP value of each feature is calculated and the prediction made is compared to the baseline prediction to find out how the feature affects the prediction.

6.10 HCI and XAI Systems in Different Real-life Sectors

A basic machine interface relies on the many inputs it receives. It communicates with the human through the interface using one or many modes of inputs and outputs. Systems usually are sensor based, visual based, audio based, or a combination of them. HCI in real-life sectors are as follows:

Eye tracking:

When users track someone's eye movements, users can figure out where they look. The whole idea of eye tracking is to track a person's point of gaze and use it as an input for machines to do their jobs. Users look at things from a certain point of view, which can be the screen of a computer or a camera lens. Another example is the working of camera software in our smartphones. Eye tracking will also help clerical workers and receptionists to decrease distractions. It helps in fine arts and sports too. The data of where an expert or professional is looking and how their vision helps them perform better can be used to educate the new and upcoming players.

Speech recognition:

Speech recognition is also another aspect of technology these days, which is directly connected with the concept of human–computer interaction. Essentially, speech recognition is software that is developed to understand human speech and take it as input. Virtual assistants like Google Home, Amazon Alexa, and Siri use speech recognition technology to function. Even though it is not perfect, it can improve with the development in AI. When perfected, it will completely make obsolete the work of typing. The typing speed of an average person is 40 words per minute, while he can speak easily up to 150 words per minute. This shows us that the speaking rate is comfortably more than three times that of the typing rate and speech recognition can, therefore, easily triple the productivity of tasks that predominantly depend on the typing skills of the user.

Internet of Things:

IoT applications are now almost everywhere in our daily lives and almost every business. IoT connects devices together and lets them talk to each other so that menial tasks can be done automatically and in a way that is more modern [7–10]. IoT collects data that businesses can use to make important decisions about how to boost their profits. In other words, IoT devices are not just for collecting data. The use of this concept is all over the place, and it can be used in many different ways. People who use IoT as a base can then use other technology to grow. These devices can also help people be more productive by letting them talk to each other. Time management is very important, and it can have a big impact on how well users do. It can help users do better work, make better decisions, stop procrastinating, and lessen stress and anxiety. People who use IoT devices can use them to help them manage their time better. For example, smartwatches can have apps like scheduling, taking notes, and reminders that can help people remember small things, manage their time, and be more productive. They can also communicate these apps to other devices.

Cloud computing:

People use cloud computing to store and manage data on servers and databases that can be accessed over the Internet. It needs resources like storage space and computing power when they are needed, and there is no way for users to control them. Twenty to twenty-five years from now, things will be very different at work. Businesses are now moving toward becoming remote workplaces that use cloud computing. Employees do not have to go to work every day, and they can work from home. This saves money for the company by not having to build and maintain a workplace. It also saves time for the employees because they do not have to drive to work. Workers do not think they need to be in an office in order to be productive, according to a study. The very same study concluded that 43% of workers think they would be more productive if they could work from home more often. Employees can work from home through the Internet and still be as productive as they would be if they worked in the office. With cloud computation and mobile devices that provide access to cloud-based applications and management work, workers can easily prefer the online work from home option. Cloud-based SaaS (Software as a Service) applications make it easy for people to work together and speed up their work. Any user with the right permissions can access the data on the cloud from anywhere in the world with an internet connection and talk to other employees in real time. These technologies also make it easier for teams to work together and communicate with each other from afar, which increases productivity. The future workforce will shift from being location-based to work-based because of SaaS and human–computer interaction; so we can be sure that this will happen [11–13].

AR/VR:

Immersive technologies like AR and VR connect humans with the digital world so that they can get help with their daily tasks and be more productive. Employees can interact with computers and their environment without having to use their hands. Smart glasses let them do this. When these glasses are combined with other interesting technologies, like cloud-based services and mobile apps, they can give users important and valuable information while they are doing their jobs. In the last few years, the development of smart glass technology has been very fast.

Features like see-through displays, voice controls, and HD cameras have not been added to the smart glasses. They are now being sold by many different companies. And this progress in this technology has made it easier for people to work with computers. They also let people learn on the job. For example, a chef who uses this kind of technology can learn new recipes quickly and start making them at the same time. This helps cut down on the amount of time there is between songs.

People who work with big machines also use AR and VR apps. Cameras with augmented reality software can be used to show a diagram of how the machines work. AR and VR can do a lot for the education business. Students can use augmented reality in the classroom to help them visualize theoretical ideas and make what they learn from books more real. AR and VR help workers improve their precision, speed, quality of work, and make sure that the job is done right the first time. Thus, it cuts down on the amount of time needed for rework or correction and gives workers more time to work on other things that need their attention.

Research in human–computer interaction (HCI) is focused on things like user personalization, behavior and sentiment analysis, and how humans and computers communicate with each other (neural-link). When researchers and developers come up with new ideas in the above areas, they can make HCI tools that can be changed to meet the needs of the people who use them. This makes them more productive. There are ways that human–computer interaction technology can be made to help people who are not very good at doing their accounts.

XAI Systems in Real-life Sectors

Data scientists all around the world have realized the need of explainable AI and have built systems that push the effect of the field into the real world. XAI is being used in high stake industries where the predictions made by AI must be understood thoroughly to ensure its accuracy and to prevent a loss in terms of lives, money, etc.

Healthcare:

The potential of AI in the healthcare industry is very high and the subsequent risk is just as high if not higher when merged with an untrustworthy AI. It is plain to see the consequences of a misclassified disease or an error in medical imaging. But that problem is dealt with when the AI system explains its prediction to the doctor. Right now, AI is being used to improve numerous tasks, including decision-making, predictions, risk management, and even diagnosis by scanning medical images to identify anomalies and patterns undetectable by the human eye. When the cost of a mistake is high, such as malignant tumor misclassification, it leads to unnecessary financial costs, elevated health risks, and personal trauma, which are avoided by explaining the prediction and classifying ailments correctly.

Banking and financial services:

XAI can revolutionize this industry. It has already started to leave its mark. Credit risk assessment, premium estimations, and prediction of stock prices are all avenues where AI is already being used in the industry. This is only improved by the introduction and integration of XAI. The uncertainty of AI predictions is now removed, which provides an assurance to the money spent. In this regard, Apple recently made headlines for allegations that its Apple Card uses biased algorithms to set credit limits and as a result discriminates against women who apply for the card. It was even claimed that men who apply for the card receive between 10 and 20 times higher credit limits than their wives. The outcry led to reputational damage and an investigation by the New York State Department of Financial Services.

Explainable AI not only provides the ability to predict fraud, but it can also tell you why XAI can determine what specific factors are associated with every fraud transaction prediction. Traditional AI analytic systems constantly struggle to give consumers an explanation for their credit decisions and XAI supports the critical need while giving the decision factors. XAI explains new patterns and trends emerging in high-risk areas.

Automobiles:

The entire autonomous driving industry relies on trustworthy AI and their ability to correctly make key decisions. This is only possible when the decisions are explained, and it is the key to understanding the capabilities and limitations of the system before its deployment. XAI has been used in high-risk decisions like braking assistance, automatic lane switching, and full autonomous driving along with low-risk operations like self-parking and voice assistants.

Marketing:

XAI is used to predict what a customer wants and, more precisely, why they want it. It is used for more personalized recommendation system. With the use of XAI, we increase conversion rates and basket sizes and reduce abandoned shopping carts. XAI contextualizes ads and provide relevance to product promotions leading to higher conversion rates.

Crime and fraud:

XAI can predict and reveal what factors are associated with theft and reduce false positives, enable additional useful verifications, and reveal emerging trends. It helps to detect fraudulent credit schemes by predicting and explaining why someone should not receive credit.

6.11 Shortcomings, Goals/Future Research Opportunities of HCI and XAI

6.11.1 Shortcomings of HCI and XAI

HCI is an ever-growing field, and it is far from perfect at the moment. It will also change with the growing trends and developing technology. Some of the problems that we face with HCI at the moment are:

- Some HCI applications require users to learn how to operate the application. This is very counterintuitive as the point of HCI, and automation is to make the use of machines easier.
- HCI mainly depends on design and a good GUI. Therefore, all problems with GUI system carry over to here as well. Problems like boot failure and monitoring/maintenance issues would occur.
- The world is made up of many languages and dialects. Even English is spoken in so many accents. It is almost impossible to make machines perfectly recognize all accents and dialects of all languages. It is even harder for machines to understand our grammar. As per the Winograd Schema, computers would not understand basic pronouns when there is ambiguity that is present when context is missing.

6.11.2 Shortcomings of XAI

Explainability is both interesting to think about and important to society. Users can use XAI, which is a very powerful tool, to get a lot more information than you can with traditional linear models. A lot of things go wrong with XAI, such as the following.

Confidentiality:

There are times when an algorithm is private, a trade secret, or it would be a bad idea to show it to other people, because there were flaws in the training information, model, or goal function that it used to figure out how to do what it did. The AI system has not learned how to see the world in a biased way.

Unreasonableness:

A lot of normal, legitimate data is used to make decisions that are not reasonable, biased, or not in line with what other people would think. It is important to look at how an AI system came up with its decisions so that you can be sure that they are fair. Note that there are two ways which decide reason-ability: One is from the point of view of AI algorithms, which use specific information to make decisions.

Injustice:

The system must make sure that it is in line with the law and moral code. Failing to abide by the legal or moral codes can prove extremely catastrophic to humanity.

6.11.3 Goals/future research opportunities of HCI and XAI

HCI wants to make systems that are functional, usable, and safe. To do this, researchers and developers need to know the things that make people use technology the way they do. They should try to come up with techniques and tools that make it easier to build systems that work well, efficiently, and safely for humans and computers to interact with each other. Designers must always put the needs of the people first, and the technology must be in line with what the people want. HCI is based on the idea that people who use a computer system should come first. People's needs, abilities, and preferences for how to do different things must be taken into account by the people who make computer programmers. People should not be put in a situation where they are forced to change the way they use and interact with a system just to fit in. Instead, the system should be modified and made flexible accordingly.

People should not have to change how they use a system in order to fit in with it. Instead, the system should be made to meet their needs.

- People who study human–computer interaction pay a lot of attention to these two things.
- Multiple ways to design new computer interfaces so that they are efficient in how quickly they can be learned, how quickly they can be found, and how easy they are to use.
- Methods for making things work, like by using software.
- There are ways to look at how usable and other features interfaces are and how well they work together.
- Methods for watching and learning about how people use computers and what that means for the rest of society.
- Methods to figure out if someone is human or not.
- Models and theories about how people use computers, as well as ideas about how to design computer interfaces, like cognitivist user models and activity theory [13–15].
- Perspectives that think critically about the values that underlie computer design, computer use, and HCI research [14].

HCI and Industry 4.0:

With the rise of industrial automation and the interconnection of devices, there are new problems that need to be solved. Humans are being pushed out of the workplace and replaced by machines. People who study human–computer interaction (HCI) try to improve and help people in real ways. This brings together the power and efficiency of automated machines and the flexibility of human intelligence. The goal is to gather, combine, and show scientific excellence from a variety of fields while meeting the needs of businesses in a lively way. Scientists and researchers in the fields of manufacturing, production management, computer science, psychology, and many more are needed to help with the human–computer interface at Industry 4.0. They should help with their scientific contributions, as well as give interesting and insightful research advice.

Industry 4.0 refers to a new way of thinking about how people and machines work together. The idea is based on smart and connected systems that can easily control the process of industrial manufacturing. As more and more machines make their own decisions and interact with production, the integration of humans into the process has to be done in a different way. We look at what is going on in the world of human–computer interaction in the field of the hour, which is called Industry 4.0. We focus on new ways of doing things that use augmented and virtual reality technology.

HCI and AI:

Early research has laid the groundwork for both AI and HCI. Both fields have a lot in common. It has been a while now, but there were a lot of close connections between the fields at first. Artificial intelligence has been changed by technology like deep learning that has come out very quickly. This has led to a split in practical methods and tools for things that do not have to do with artificial intelligence. In particular, new artificial intelligence techniques now offer new ways for humans and machines to interact, as we talked about in the past.

HCI and automation:

In the modern world, technological advancements and social living call for more technology to be used and for an interaction platform to be automated. The only goal is to cut down on the amount of labor or human effort and maximize the amount of work that can be done. The next generation wants to see these kinds of changes in current systems, but no one can say what kind of changes will be made in future models. For example, to meet these needs, an

HCI future framework that can be used to automate interaction and interfaces must be built.

When machines learn, they need to make it easy for humans to understand and evaluate their predictions or decisions. This is important for AI that is transparent, accountable, and trustable. Many people and businesses could be affected by these systems, which are called explainable AI (XAI). They can help people understand how AI works. There are a lot of ways that they could help humans and AI work better together in areas like medicine, finance, the legal system, autonomous cars, or the defense. Some other people have said that XAI could help with bias and other problems that could happen when computers make decisions for us instead of us making them ourselves instead of having them. Research by PAI shows that the explainability techniques that are being used to make AI more transparent and accountable for users and other outside groups are not yet up to the task of making AI more transparent and accountable for them. A paper called "Explainable Machine Learning in Deployment" written by Umang Bhatt and Alice Xiang was the first to look at how ML explainability techniques are being used in the real world. Many people who use explainability were interviewed by PAI. They found that in its current form, XAI is best used by engineers and developers to find and fix mistakes in their models, not by people who do not use the system to explain things to them. There is a big difference between how easy it is to explain things in practice and how easy it is to be transparent, because the explanations are mostly for people inside the company. Gaps like this one are very important when it comes to how AI is ruled. A company called PAI has found that more changes are needed to XAI techniques in order for them to work as intended and help people outside of the company understand and evaluate automated decisions. This includes people like end users, policymakers, and people from other companies.

HCI and Usability

In HCI, usability is a very important thing. It tries to make systems that are easy to learn and use. Many systems and products that we use every day seem to be made with little thought for how people will use them. Because of this, there is a lot of wasted time, frustration, and mistakes that could have been avoided. They include things like computers and remote controls and phones. Some of these things can be very bad at times and cause a lot of people to be unhappy.

Consider the case of a photocopying machine. It is likely to have buttons that are meant for generating the copies, clearing the settings, etc. The button that performs the former function is the one with a diamond-like symbol, which is majorly used globally. Even then, it might seem confusing to a user who is new to this. A usable system is one that is easy to learn and easy to remember how to use. A system must be effective and efficient in doing the job with regard to all the constraints of the user such as budget, time, staff, etc. The system must be safe to use and must not be unethical, immoral, or illegal in any way. The system must be enjoyable to use. This is done by the UX designers.

Norman's model of interaction (sometimes called the execution–evaluation cycle) is a notable model for HCI. It proposes that a user first establishes a goal and then performs actions using the system to achieve that goal. A system then reflects the output of those actions on the interface. A user observes the interface and evaluates if their goal has been met. If not, a new goal is established, and the cycle is repeated.

Disciplines contributing to HCI:

As discussed earlier, the field of HCI is a tapestry over a wide range of topics, and its progression has heavily relied on contributions from many other disciplines. Some of the main disciplines that have contributed to HCI are:

- **Computer science:** This is fairly simple to understand. The technology aspects towards with software designing, development, and maintenance is too less. The more focus is to make systems more effective and computable. Systems like user interface management systems and user interface development environments are from the computer science fields exclusively. There are several prototype models/ tools to build such effective system. On other side, Graphic development and user interface are focused to design/ make the system more appealing and engaging to end users.
- **Cognitive psychology:** This field contributes greatly to HCI. The study of the capacity of information processing of a user. The capabilities and limitations of users based on criteria like age, education, etc.
- **Social Psychology:** The social and organizational structures of the user play a big part on their interactions in daily life. This aspect is very important in HCI to make computers accessible for everyone.
- **Ergonomics/human factors:** These are other factors like hardware design and display readability.

- **Linguistics**: Linguistic differences between people of different parts of the world should not be an issue with HCI. So, natural language interfaces are developed.
- **Artificial intelligence:** This up-and-coming field is the future of technology and intelligent software is what needs to be developed now.
- **HCI and UX fields:** HCI was earlier considered academic and user experience design-centric [5]. The growth and importance of HCI and UX have clearly expanded, requiring prowess from an ever-growing list of other disciplines.

6.12 Evaluation Methods for HCI and XAI

Evaluation of HCI:

Different HCI design approaches have been developed to create user-friendly, efficient, and intuitive systems. HCI systems are studied and refined by evaluation and have formed the four different approaches to design them. The four main approaches have become popular, namely cognitive approach, anthropomorphic approach, predictive modeling approach, and the empirical approach.

Cognitive approach:

The cognitive approach considers the human brain and the sensory perception system of humans to develop their user interface. All design elements take inspiration from how the human brain functions and deeply absorbs from the field of cognitive psychology.

Anthropomorphic approach:

This approach has systems built in human-like procedure and mirrors human–human interaction to make it easier to use. Interface error messaging is often written this way, such as, "We're sorry, but that page cannot be found." Another example is the use of avatars in computer-based automation, as can be found in automated telephony systems.

Predictive modeling approach:

This is based on human information processing theory and works in terms of time taken to efficiently complete the set task. For example, the average time taken for a human to load a web page, look at the contents, cognitively process what is being seen, etc. is all taken into account when creating this model.

Empirical approach:

The empirical approach to HCI is useful for examining and comparing the usability of multiple conceptual designs. In addition to a qualitative assessment of user preferences for a conceptual design, measuring users' task performance is important for determining how intuitive and user-friendly a web page is.

Evaluation of XAI:

In the current XAI structure, the accuracy of the model interpretability is performed with human intervention and supervision. The problem is that human measurements are subjective, tedious, time-consuming, and are prone to inaccurate evaluations. The community has recently proposed different metrics other than evaluating the correctness of XAI methods, such as persuasibility, fidelity, and generalizability, to evaluate XAI properties [6]. The metrics are often domain-specific or very conceptual. Going forward, evaluating the correctness of models must also be automated. Automated XAI model evaluations do exist and work by perturbing results in a specific region, but this violates the assumption that test data and training data come from the same distribution. XAI must be evaluated in such a way that the process is automated fully and the evaluation metrics for evaluating the correctness of models are objective.

6.13 Ethical, Legal, and Social Issues of HCI and XAI

The ethical guidelines for XAI are divided and organized into three levels of abstraction [7]:

- The principle guidelines put forth by AI algorithms must respect and abide by the human anatomy, prevent any sort of consequences, and be fair.
- Guidance in the form of seven major requirements, which illustrate how AI solutions are developed, is based on transparency and fairness and correlate to XAI.
- A list of questions that can be useful in developing the reliability and loyalty of the AI solution.

It is essential for the users and developers of AI solutions to comprehend and be aware of these guidelines and norms. They will require AI solutions that have:

- The functionality to trace down a dataset and the systems decision in terms of documentation.
- The ability to describe the technical procedures of the AI systems to humans.
- The functionality to put forth a necessary explanation on the decision-making process when the system has considerable impact on the lives of people.
- No bias or leniency in the dataset, exploitation of human leniency, or unfair competition.
- Implemented a user-focused design process with the help of stakeholders all throughout the development of the system, starting from the analysis phase and proceeding with the regular feedbacks for the system implementation.

Ethics in human–computer interaction (HCI) covers a large range of topics, which include privacy, ownership and property, human welfare, universal usability, trust, autonomy, freedom from bias, informed consent, and accountability, among others. HCI is deeply concerned with the ethical implications of technology design and use.

6.14 Conclusion

Artificial intelligence is undoubtedly going to play a huge role in the future and HCI and XAI are important aspects of artificial intelligence. To build any useful systems, both these areas of research must be considered heavily. We introduce both topics in this chapter and delve into the many areas of interest. In the near future, artificial intelligence will be used to remove the burden/work load from the shoulders of skilled engineers in this field (useful for IoT-based cloud applications from agriculture to transportation) and to deal with these issues the right and accurate way. On another side, skilling too many people for implementing AI in near future is a challenging task and time-consuming (lengthy) process.

References

[1] J. R. Anderson. "The Architecture of Cognition." Harvard University Press, Cambridge, Massachusetts, 1983.

[2] G. Abowd. Agents: "Recognition and Interaction Models." In D. Diaper, D. Gilmore, G. Cockton, and B. Shackel, editors, Human-Computer Interaction - Proceedings INTERACT'90, pages 143-146. North-Holland, Amsterdam, 1990.

[3] R. Beale and J. Finlay, editors. Neural Networks and Pattern Recognition in Human-Computer Interaction. Ellis Horwood, Chichester, 1992

[4] H. Alexander. Formally-based Tools and Techniques for Human-Computer Dialogues. Ellis Horwood, Chichester, 1987.

[5] R. W. Bailey. Human Performance Engineering: A Guide for System Designers. Prentice Hall, Englewood Cliffs, New Jersey, 1982.

[6] Yi-Shan Lin, Wen-Chuan Lee, Z. Berkay Celik. What Do You See? Evaluation of Explainable Artificial Intelligence (XAI) Interpretability through Neural Backdoors.

[7] Ethics Guidelines for Trustworthy AI. Report.

[8] M. M. Nair, A. K. Tyagi and N. Sreenath, "The Future with Industry 4.0 at the Core of Society 5.0: Open Issues, Future Opportunities and Challenges," 2021 International Conference on Computer Communication and Informatics (ICCCI), 2021, pp. 1-7, doi:10.1109/ICCCI50826 .2021.9402498.

[9] Tyagi A.K., Fernandez T.F., Mishra S., Kumari S. (2021) Intelligent Automation Systems at the Core of Industry 4.0. In: Abraham A., Piuri V., Gandhi N., Siarry P., Kaklauskas A., Madureira A. (eds) Intelligent Systems Design and Applications. ISDA 2020. Advances in Intelligent Systems and Computing, vol 1351. Springer, Cham. https://doi.org/10.1007/978-3-030-71187-0_1

[10] Goyal, Deepti & Tyagi, Amit. (2020). A Look at Top 35 Problems in the Computer Science Field for the Next Decade. 10.1201/9781003052098-40.

[11] Amit Kumar Tyagi, Dr. Meenu Gupta, Aswathy SU, Chetanya Ved, "Healthcare Solutions for Smart Era: An Useful Explanation from User's Perspective", in the Book "Recent Trends in Blockchain for Information Systems Security and Privacy", CRC Press, 2021.

[12] Varsha R., Nair S.M., Tyagi A.K., Aswathy S.U., RadhaKrishnan R. (2021) The Future with Advanced Analytics: A Sequential Analysis of the Disruptive Technology's Scope. In: Abraham A., Hanne T., Castillo O., Gandhi N., Nogueira Rios T., Hong TP. (eds) Hybrid Intelligent Systems. HIS 2020. Advances in Intelligent Systems and Computing, vol 1375. Springer, Cham. https://doi.org/10.1007/978-3-030-73050-5_56

[13] Tyagi, Amit Kumar; Nair, Meghna Manoj; Niladhuri, Sreenath, Abraham, Ajith, "Security, Privacy Research issues in Various Computing Platforms: A Survey and the Road Ahead", Journal of Information Assurance & Security. 2020, Vol. 15 Issue 1, p1-16. 16p. Add here:

[14] Madhav, A.V.S., Tyagi, A.K. (2023). Explainable Artificial Intelligence (XAI): Connecting Artificial Decision-Making and Human Trust in Autonomous Vehicles. In: Singh, P.K., Wierzchoń, S.T., Tanwar, S., Rodrigues, J.J.P.C., Ganzha, M. (eds) Proceedings of Third International Conference on Computing, Communications, and Cyber-Security. Lecture Notes in Networks and Systems, vol 421. Springer, Singapore. https://doi.org/10.1007/978-981-19-1142-2_10

[15] Tyagi, A. K. (Ed.). (2021). Multimedia and Sensory Input for Augmented, Mixed, and Virtual Reality. IGI Global. http://doi:10.4018/978-1-7998-4703-8

7

BMRP: Blockchain-Based Medication Reconciliation Process in EHR to Improve Patient Safety

S. Satheesh Kumar[1], Manjula Sanjay Koti[2], V. Vinoth Kumar[3], V. Muthukumaran[4], and B. Swapna[5]

[1]Department of Computer Science, School of Applied Sciences, REVA University, India
[2]Department of MCA Dayananda Sagar Academy of Technology & Management, Bangalore, India
[3]School of Computer Science Engineering and Information Systems, Vellore Institute of technology, Vellore, India
[4]Department of Mathematics, College of Engineering and Technology, SRM Institute of Science and Technology, Kattankulathur - 603203, Tamilnadu, India
[5]Department of Electronics and Communication Engineering, Dr.M.G.R. Educational & Research Institute, India
Email: satheesh.sampath@gmail.com; manjula.dsce@gmail.com; drvinothkumar03@gmail.com; muthu.v2404@gmail.com; swapna.eee@drmgrdu.ac.in

Abstract

For a long time, blockchain has been a fascinating research topic, and a variety of businesses can utilize it because of security, privacy, confidentiality, and decentralization. Blockchain technology has a lot of potential in the healthcare sector. Despite this, there are issues with data reconciliation, integrity, and management in electronic health record (EHR) systems. In a formal medication process, reconciliation in healthcare provides collectively working with patients, families, and care providers to confirm the precious and complete medication information, which will communicate consistently among all phases of care. The medication reconciliation process carries a review of all the medication that a patient takes. The drug can be included, altered, or may be discontinued. One of the components of medication reconciliation is medication management, where this management gives information to the prescribers in arriving at suitable prescribing decisions for the patients. The leading cause for the adverse drug event is when there is a change in the frequency, and the communication problems between the various care settings play a significant role in their occurrences. Data mining adopts explorative data analysis to understand the frequency of medication events in the electronic health record (EHR) data to perform the description analysis. Explorative data analysis is a quantitative technique used to understand the data when a statistical hypothesis exists. The various applications of process mining such as discovery, conformance, and enhancement of process can be applied to analyze an enormous amount of data concerning patients and medication events in EHR. Accordingly, this data generates a description for the structured approach, which in turn can be used to discover adverse drug events. To identify the adverse drug event, the following steps need to be taken care of: (1) creation of best possible medication history (BPMH) by taking into account the patient name, dosage, root, and frequency; (2) comparison of best possible medication history (BPMH) with admission, transfer, or discharge medication records; (3) any changes in medication order to be documented and communicated, thereby ensuring patient safety. By adopting these techniques, we can create an automation system to detect adverse medication in the health records. Thereby, physicians can prescribe the proper medication for the patients. In this chapter, we have taken the EHR-based medication reconciliation problem for post-surgery patients. We have used the R tool to estimate the adverse drug effects, and the result indicates that the physician takes care of these adverse drug effects, thereby ensuring patient safety. In this chapter, we address how the blockchain technology can be applied to change EHR systems and how it might be a solution to issues with prescription reconciliation. The primary objective of our suggested framework is to use blockchain technology for EHR before providing medication reconciliation in electronic records to increase patient safety.

Keywords: Blockchain, EHR, Medication Reconciliation, Best Possible Medication History.

7.1 Introduction

An abundant wealth of data is available in hospitals, and this data gives information about the healthcare process, known as event logs. Process mining utilizes the inputs from these event logs to discover, monitor, and improve the actual process. EHR stores patient history information and retrieval of data from EHR is easy compared to paper-based medication records [1]. To ensure patent safety in healthcare, the implementation of EHR plays a significant role. If proper care is not taken in implementing the EHR process, this can lead to unforeseen death due to a lack of error handling in the clinical care process. The mistakes in EHR can be categorized in three ways: (1) data fragmentation, (2) disruptive performance, and (3) negative impact on workflow issues. Identifying the EHR log clinical task data is a very tedious task. Since this helps in identifying all possible unforeseen events [2]. The two significant challenges include (1) detection and investigation of error events and (2) identification of system issues, which leads to the monitoring of error events. Sometimes clinicians fail to make an entry in the EHR, thereby encountering for data entry errors; hence, there is a need for standardizing the medication reconciliation process, which in turn can be used to reduce the number of patients with unforeseen discrepancies, thereby reducing the occurrences of medication errors [3]. Text mining is the method that can be used to extract process-related information from the process log. Health information is stored as a record in the medical field that is processed using process mining techniques to provide quality services and reduce costs. Finally, exploratory data analysis and text mining methods can give the automated system to provide patient safety in healthcare [4]. The primary importance of cloud-based EHR is not so expensive and more scalable, thereby protecting the privacy of the data. Clinicians use Clinical Decision Support System (CDSS) software to make efficient decisions concerning particular practices and multiple management systems. This software records the patient condition and alerts about the patient issues [5]. A formal medication reconciliation process compares the patient medication orders with all the other medications the patient was taking. Medication errors such as drug interaction, dosing errors, duplications, and omissions can be avoided. Wherever there is a new medication or the existing medication is rewritten at every stage of these transactions, the reconciliation process will be taken care of. The best medication reconciliation utilizes the best possible reconciliation history (BPMH) to estimate the majority of unaware discrepancies that lead to medication errors. These medication errors are originated while acquiring patient medication history. General approximators, also known as simple learning algorithms, are categorized as predictive modeling. General approximators predict the significant outcomes by considering the relationship in the data, which are applied to extract the repeated and consistent patterns

[6]. Recent developments in technology change how we use and perceive things and have an impact on many facets of our existence. Technology is discovering new ways to improve the healthcare sector, just as it has in other aspects of life. Security enhancements, user convenience, and other facets of the healthcare sector are some of the main benefits of technology development. Systems for electronic medical records (EMRs) and electronic health records (EHR) offered these benefits. However, they do have certain worries regarding the safety of medical records, user ownership of data, data integrity, and other issues. Using cutting-edge technology like blockchain could be the solution to these issues. This chapter is organized as follows. Section 7.1 deals with an introduction to the topic of discussion. Section 7.2 presents work related to the research undertaken. Section 7.3 presents the novel approach of blockchain technology with EHR medication reconciliation, and Section 7.4 elaborates the medication reconciliation process. Section 7.5 explains in detail the best possible medication history. Section 7.6 deals with mitral valve replacement of patient condition, and Section 7.7 deals with dataset description. Section 7.8 explains the methodology. Section 7.9 presents experiments and results. Finally, Section 7.10 offers a conclusion. However, a thorough literature survey on the research topic has been done, and nowhere in the literature such a kind of study existed.

7.2 Related Work

In [7], the authors examined the extraction of past clinical data and occasion logs of medical care data frameworks. This technique manages the trouble of certifiable clinical pathways and date stamp occasion log divisions. Process mining strategies empower the obtaining of essential bits of knowledge into clinical pathways dependent on hints of clinical pathway examples – this technique is fit to examine unfavorable occasions.

In [8], the authors deal with the massive volume of data management and record of patient treatment in hospitals. Process mining can manage and view the enormous volume of data with execution using classification techniques. The significant challenges and different types of event data are analyzed in health information systems (HISs).

In [9], the authors presented the detection of medication events and medication revisions using machine learning classifiers such as probabilistic classifiers and decision trees. To detect and improve patient safety, process mining can help identify the medication error events resulting from identifying uncovered factors in medication revisions.

In [10], the authors discussed the process mining techniques, perspectives, and tools with its methodology in the medical field. They identify emerging technology and trends in data mining for healthcare information systems. Discussion helps researchers to choose models and algorithms for applying in their applications to analyze healthcare data in EHR.

In [11], the authors deal with the posterior analysis techniques, which can be used to record event logs in information systems so that process information can understand and design the high-quality process for the patients to provide treatments and ensure patient safety.

In [[12], the authors have discussed the uses of predictive model and flexibility of process mining techniques and how it allows users to do the conformance checking and analyzing deviations for diagnostics. This approach can be applied in various departments to get feasible results using the ProM toolset. It declares the valuable insights related to process conformance.

In [13], the authors presented the EHR-based patient care processes and workflow usability. Also, they discussed patient behavior detection of deviations from indented process models to reduce medical errors and increase patient safety. The suggestion has been given to enhance healthcare process efficiency and patient satisfaction, which can increase EHR adoption. Identification of best practices and coordination of care among clinical staff will improve the efficiency of the EHR usability.

In [14], the authors deal with the confidentiality of patient data and how it can be protected with cryptosystems using access control aware search technique (ACAS) with AdvancedMD EHR software inpatient health information from unauthorized access. Cloud-based encryption and decryption techniques are used in cloud-based EHR access to extract patient data, giving secure access to physicians and patients without discrepancies.

In [15], the authors have presented the process of medication reconciliation to create accurate patient medications. This can be compared with existing patient medication history to avoid medication errors such as drug interactions, duplications, and dosing errors. Existing orders are rewritten when the new medication reconciliation process is done at every transaction.

In [23], according to the authors, blockchain is a network of blocks that grows exponentially as transactions are stored on them. The distributed methodology used by this platform ensures that each distributed piece of data, also known as data, has shared ownership while allowing for data sharing. Blockchains are peer-to-peer network-managed distributed ledgers that hold batches of hashed transactions. Data security, privacy, and integrity are just a few benefits of using a blockchain, even without a third party.

7.3 Background

Blockchain technology can enhance healthcare by putting the patient at the center of the system and enhancing the security, privacy, and interoperability of health data. This technology might create a new framework for the interchange of health information by making electronic health records (EHRs) more effective and secure (HIE). For diagnosis and treatment in healthcare, electronic health records (EHRs) include vital and highly sensitive personal data. These records are a useful source of patient data. The sharing of health records is necessary to make the healthcare system brighter and improve service quality. An EHR is an electronically stored structure of a patient's health data developed and maintained throughout the patient's life. It is often held by and distributed among different hospitals, clinics, and health providers. Blockchain, which enables a sharing and security mechanism, can be a future solution for data sharing in health and precision medicine. The chapter's main contribution is a systematic literature study that summarizes previous studies on EHRs and blockchain; this research focuses on how a blockchain structure can be used in healthcare for EHR storage and retrieval control. The blockchain architecture is shown in Figure 7.1, which shows the entire process of an effective communication system and activity to the blockchain network.

The term "blockchain" refers to a distributed ledger system that was first associated with the cryptocurrency Bitcoin. Using public-key cryptography, it generates an append-only, immutable, and time-stamped chain of material. The blockchain paradigm was initially designed to maintain a financial ledger; however, it may be extended to give a general framework that incorporates distributed computing resources in any industry, including healthcare. In conclusion, this study aims to research and evaluate the literature on blockchain in healthcare. Furthermore, we attempt to uncover obstacles and open questions from the survey into possible solutions for healthcare data concerns, such as storage and privacy concerns. Finally, we also present a taxonomy to help people better comprehend the blockchain ecosystem explained in Figure 7.1 and to aid in future research.

Figure 7.1 illustrates the process of the novel approach, blockchain-based electronic healthcare record medication reconciliation process.

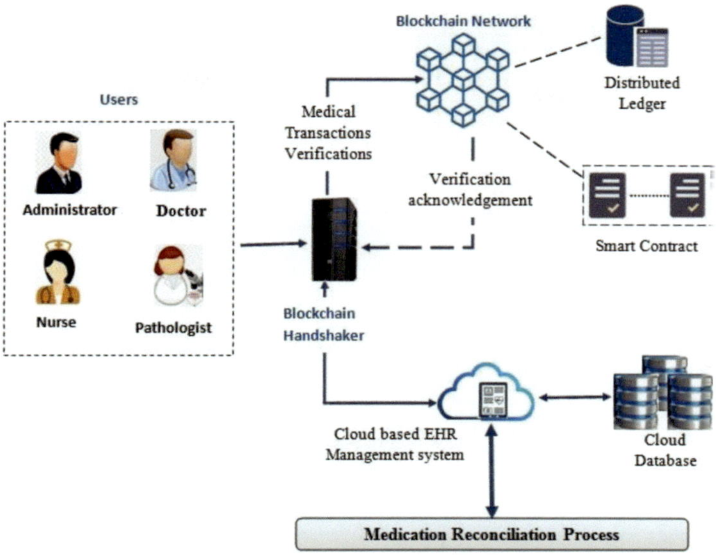

Figure 7.1 Architecture framework for blockchain-based EHR-medication reconciliation process.

7.4 Medication Reconciliation

Medication reconciliation is a process to create accurate and possible medication for the patients. In this, the current drug can be compared with the existing medication orders that the patient has been taking. Implementation of this process can be used to find the majority of unintentional discrepancies, which can lead to medication errors in a patient's medication history, and also it reduces the adverse drug event (ADE) when the system is in computer physician order entry [16]. Additionally, prescription errors such as drug interactions, dosing mistakes, duplications, and omissions can be prevented by using the medication reconciliation process. Prescription errors can be minimized by accurately recording medications taken in the clinic, ensuring clinically required home medication dosages are correctly documented, and clearly identifying which home medications need to be continued or discontinued after discharge. The establishment for remedy compromise is the best medication history (BPMH). It contrasts from a typical medication history that includes (1) an intentional gathering process with the patient (or family) and (2) an overview of, without a doubt, one more trustworthy wellspring of information (e.g., an ordinary medication database, an assessment of remedy vials, or contact with a medically trained professional). Solidifying a helpless

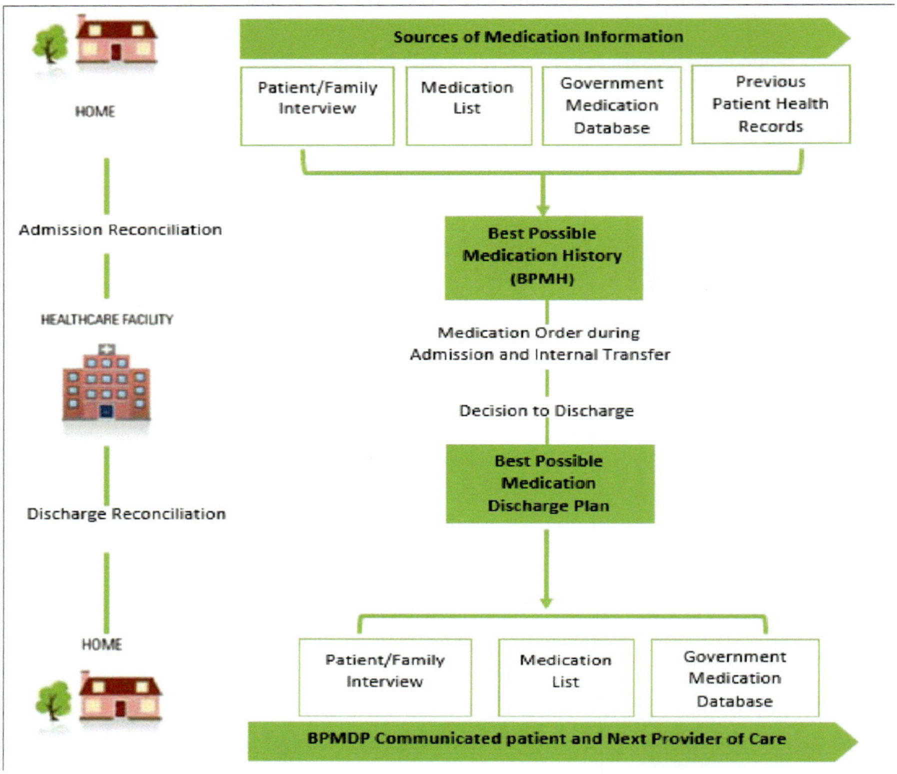

Figure 7.2 Overview of the medication reconciliation process.

medicine history with solution orders will not further develop treatment and may even "hard-wire" coincidental inconsistencies. At long last, acquiring a BPMH has various obstacles, including new patients to their medications. Among clinics and surprisingly clinical settings, the amount and seriousness of medicine compromise activities may sensibly vary. The medication reconciliation process is explained in Figure 7.2.

It is debatable whether clinics should prioritize internal exchanges, releases, or confirmations. Some studies suggest that internal exchange centers present significant medication risks as they transition from concentrated care to distribution. In any case, most emergency clinics decide to zero in taking drugs compromise at the hour of affirmation or discharge. Both arrangements have execution difficulties, and observing the best methodology for a specific clinic is troublesome. At last, some kind of release compromise is needed; all things considered, the essential objective of drug compromise is to forestall unintentional prescription blunders brought about by hospitalization. Discharge compromise's principal issue is fusing it into the entire release process, which as of now requires different activities from doctors, medical attendants, drug specialists, and other wellbeing experts. It is enticing to simply hand along with a drug compromise structure or a letter with a rundown of remedy changes. Isolating medicine compromise from related release activities, for example, getting release outlines and medicines ready, may restrict its entire viability and give the incorrect impression that prescription compromise is not a clinically critical interaction that simply adds to doctors' responsibility. The entanglements of the release technique are kept away from compromise at confirmation, yet the time strain upon affirmation, particularly for truly debilitated patients, makes acquiring a BPMH troublesome. Practically speaking, emergency clinics address the issue of time requirements by characterizing "on affirmation" as "inside 2448 hours of confirmation." Eliminating the time pressure also removes the risk of drug abuse affecting the formulation of treatment orders during verification. If not integrated into a significant therapeutic exchange, such as placing orders upon confirmation, the idea of medication reconciliation might seem like pointless work done for perplexing reasons, like the release cycle. Generally speaking, starting the prescription compromise process as quickly as time permits after affirmation might save time over the long haul. Regardless of whether confirmation compromise quickly issues affirmation orders, in case of a BPMH as of now exists, the release method will be more precise and proficient. Furthermore, the Accreditation Canada sign, a characteristic inspiration, is zeroing in on permission. In particular, these ideas are unique.

The accomplishment of the Medication Reconciliation SOP execution will be impacted by the unique circumstance (or climate) in which it is applied. Outside elements like wellbeing strategy, public rules, and accreditation necessities for prescription compromise, just as inner components like the particular qualities of every medical care association, will have an effect. Perspectives like culture, initiative, the size and design of the association, the type of responsibility for mediation, and the accessibility of assets are altogether factors that may impact the adequacy of the execution and ought to be considered while making the strategy.

Figure 7.3 represents medication reconciliation transaction process steps.

Rather than simply adding extra exercises, a compelling and proficient prescription compromise system will include reconciliation of its means into existing medication for executives and patient stream activities. Subsequently,

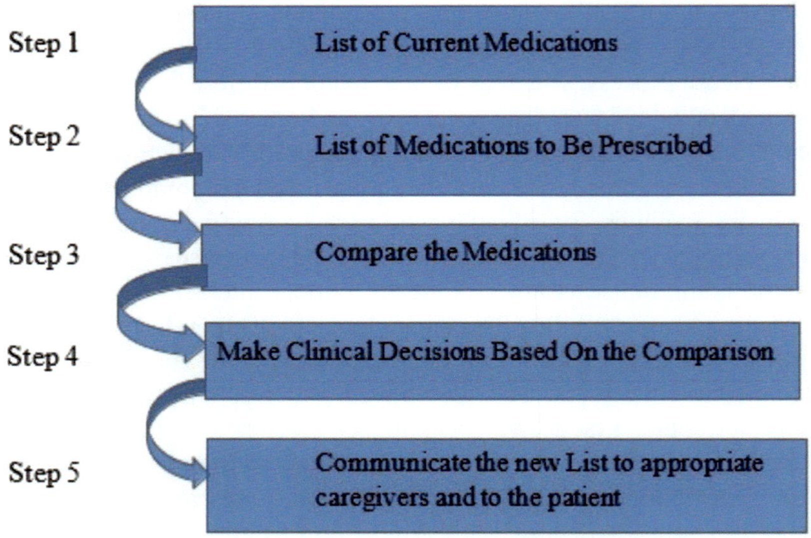

Figure 7.3 Medication reconciliation transaction process.

it is basic to perceive different spaces of patient consideration with which medicine compromise should collaborate, for example.

- First assessment of the patient
- Medication ordering, preparation, and dispensing
- Treatment record keeping
- Internal processes for patient transfer
- Clinicians exchanging information
- Discharge planning
- Education of the patient

The reconciliation process is done at every transaction when the new medication orders are rewritten. For example, the following steps include developing a medication list for a newly hospitalized patient: (1) verifying medication history, (2) documenting patient's medication history, and (3) creating medication administration records. Following the patient's discharge, the next stages involve deciding on the post-discharge drug schedule and creating discharge instructions for home medications. A three-step procedure is used for medication reconciliation in acute care: (1) compile the most thorough and accurate medication history you can (BPMH); (2) integrate the medications; (3) communicate and document [17].

7.5 Creating Best Possible Medication History

The term "best possible medication history" (BPMH) refers to a medication history that has been compiled by a clinician and includes information from a number of sources as well as a thorough and detailed account of all regular medication use (both prescribed and over-the-counter). The BPMH must include a complete list of all prescribed (based on the prescriber's advice) and non-prescribed (not based on the prescriber's advice) drugs, including:

- A prescription (medications the patient is told by the physician to take); (the prescriber did not advise the patient to take the medication)
- Medications on prescription
- Drugs sold over-the-counter.

In order to create the BPMH, a systematic strategy to collecting a medication history is necessary. This approach must include:

- Interviewing patients and family members if possible.
- Confirming and recording the past:
- If at all possible, conduct a medication interview with the patient.
- Verification of prescribed drugs with many sources, if needed, such as family or caregivers, community pharmacists, and physicians.
- Medication vials are inspected.
- Medication profiles from other facilities are listed on the patient's medication list.
- Previous patient prescription drug claim histories (Drug Profile Viewer) of Ontario Drug Benefit (ODB) beneficiaries.

Regardless of whether it contrasts based on what was really given, the BPMH incorporates the medication name, portion, recurrence, and course of drugs a patient is presently taking. For precision and proficiency, utilizing devices like a manual for gathering the BPMH might be advantageous. Different sources might be used to accumulate prescription chronicles or clarify clashing data assuming that a patient is inaccessible to partake in a drug meet. Different sources should never be utilized to supplant a potent medicine to meet with the patient and family. If conceivable, every personal medication and related dosing guidance ought to be affirmed for patients who present solution bottles or potentially a prescription rundown. Patients regularly consume drugs in manners that vary based on what is composed on the solution mark. Likewise, patients may have forgotten to refresh their own solution records with newly recommended meds. Different sources might be utilized to gather prescription accounts and explain clashing data assuming a

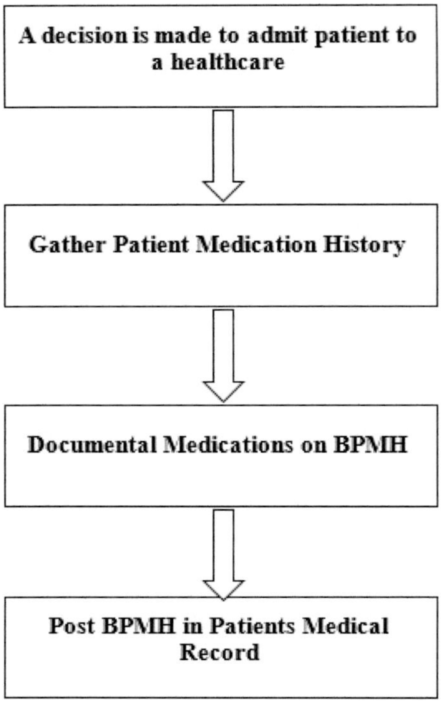

Figure 7.4 Flow map for BPMH.

patient or relative cannot take part in a medicine meet. Forestalling medicine-related mischief, frequently known as adverse drug events (ADEs), is an essential patient security concern in clinics and throughout the patient's consideration continuum. Many organizations have demonstrated that practicing medication reconciliation at all stages of patient careâĂŤadmission, transfer, and dischargeâĂŤis an effective method for preventing adverse drug events.

Medicine compromise is the most common way of assembling the most dependable rundown of all prescriptions a patient is taking, including drug name, measurement, recurrence, and course, and contrasting it with the doctor's affirmation, move, and additionally release orders fully intent on furnishing the patient with the right meds at all progress focuses inside the clinic. An ideal prescription history (BPMH) is one of the most amazing drug compromise processes that can be utilized to find inadvertent inconsistencies that might prompt medicine blunder. Prescription history can be made using an orderly course of meeting the patient/family and an audit of, without a doubt, another solid wellspring of data to confirm patients endorsed and non-recommended medicine lists. This cycle assists clinicians in seeing how BPMH is made and in getting the ideal medicine for the patients. Figure 7.4 portrays the BPMH stream map, which helps in making patient prescription records [18]. The process of best possible history flow map is defined in Figure 7.4.

7.6 Mitral Valve Replacement

An open mitral valve substitution is a technique that includes supplanting a broken mitral valve with a mechanical valve. One of the four valves of the heart is the mitral valve. It assists in the progression of blood through the heart and out to the rest of the body. The mitral valve is found between the left chamber and the left ventricle. Your mitral valve will be replaced with an artificial valve if it is not operating correctly. This ensures that blood may flow regularly into the left ventricle and out to the body without putting additional strain on the heart. The procedure is named "open" because it exposes the heart through a traditional bigger incision. The incision used in minimally invasive mitral valve replacement surgery is larger than this one. If the mitral valve is not performing properly, the operation may be required. The mitral valve can usually be repaired surgically, although it may need to be replaced in some cases. Mitral valve stenosis and regurgitation (also known as mitral insufficiency) are two different valve disorders that may require valve replacement. When the mitral valve stenosis occurs, the valve cannot fully open, allowing less blood to flow from the left atria to the left ventricle. The valve is leaking in mitral valve regurgitation. Instead of going forward into the left ventricle, some blood spills back into the left atrium. Fatigue and shortness of breath are common signs of valve difficulties. Surgery may be required if these symptoms become severe. Mitral stenosis and regurgitation can both be caused by the valve's natural aging. Mitral valve disease can also be caused by the following reasons:

- Degenerative changes
- Rheumatic fever
- Congenital heart defects
- Infective endocarditis
- Connective tissue disorders
- Degenerative conditions

The risks of mitral valve surgery vary based on the individual's medical history and general health before surgery. For example, younger and healthier people have a lesser chance of problems, whereas older people with other health issues have more significant difficulties.

The following are some of the risks:

- Consequences of the operation (such as bleeding, infection, and risks associated with anesthesia).
- The procedure may cause death. This risk varies depending on various factors, including age, heart health, and other medical issues.
- The new valve may cause blood clotting. To avoid dangerous blood clots, a mechanical valve must be treated with anticoagulant drugs for the rest of one's life.

The new valve may fail. There is a slim chance that the valve will not function. Your doctor will need to check your valve regularly to ensure it is working properly. Another valve replacement procedure is required. Artificial valves have a finite lifespan. The type of valve you have and how long you live after your first operation will determine whether you need valve surgery again. Prothrombin time (PT) results are processed to make the international standardized percentage (INR), which is used to assess how well the blood-thinning medicine (anticoagulant) warfarin is working to prevent blood clots. After substances (reagents) are acquainted with your blood test, a PT looks at how long it requires for coagulation to create. The PT is oftentimes utilized related to an incomplete thromboplastin time (PTT) to analyze the number and capacity of proteins known as coagulation factors, which are fundamental for viable blood clump development. This technique incorporates a succession of substance occasions known as the coagulation course, in which coagulation or "thickening" parts are set off individually, bringing about the development of coagulation.

Altogether, for ordinary thickening to happen, every coagulation factor should be available in proper amounts and work viably. There are two "pathways" that can trigger coagulating in a test tube during a research center test, the outward and inherent pathways. Both of these pathways accordingly unite to complete the coagulating system. Factors I (fibrinogen), II (prothrombin), V, VII, and X are completely included. The PT/INR should be possible related to a PTT, which evaluates the inborn and outward coagulating factors: XII, XI, IX, VIII, X, V, II (prothrombin), and I (fibrinogen), just as prekallikrein (PK) and high sub-atomic weight kininogen (HK). The PT is frequently estimated in a flash and contrasted with an ordinary scope of PT esteems in solid individuals. The typical results will fluctuate because the reagents used to perform the PT test vary, not only between research centers but also within the same laboratory over time. For people utilizing the anticoagulant warfarin, a WHO board was set up and suggested the utilization of the internationalized normalized ratio (INR), which is determined depending on the PT test result. Warfarin is utilized to treat various issues, including profound vein apoplexy (DVT) and some cardiovascular diseases (CVDs), like atrial fibrillation. The INR is a recipe that records for varieties in PT reagents and empowers for correlation of discoveries from various labs. At the point when a PT test is performed, most research centers report both PT and INR readings. Nonetheless, the INR ought to just be utilized by individuals who are taking the blood more slender warfarin.

7.7 Dataset Description

The dataset that has been considered for this study is from IWPC (International Warfarin Pharmacogenetics Consortium), which is publically available. In the training phase, the data points are assigned to two classes. The first class contains high required dose (HRD), the cardiac post-surgery patients who require amounts greater than 30 mg/week, and the second class has low required dose (LRD), the cardiac post-surgery patients who need doses of less than or equal to 30 mg/week. The cutoff point of 30 mg/week has been taken from the validation process, where the training and validation sets are randomly divided from the data in learning sets.

The prothrombin time (PT)/international normalized ratio (INR) is utilized to assess the extrinsic and common pathway of blood coagulation, especially factor VII assessment. In the current situation, every research center is handling PT/INR assessment for noticing oral anticoagulant (warfarin) [19].

Target international normalized ratio for PT/INR: mean 2.5, standard deviation 0.1, minimum 1.8, and maximum 3.5.

7.8 Methodology

To dispose of blunders and improve patient results, emergency clinics are under expanding strain to utilize prescription reconciliation, a deliberate method to confirm the right medicine records at confirmation, during a hospitalization, and at discharge—wellbeing-related innovation. Prescription compromise can be supported by electronic health records (EHRs), yet information quality and security are significant contemplations. Another subjective review by analysts indicated that specialized and functional

Table 7.1 Dataset description for international normalized ratio with drug dose and treatment.

Dose (mg)	Prothrombin time (PT)	Reference interval	INR	Normal/ abnormal	Treatment	Alternative treatment
3 & 4 Alternate days	43.6	**High** 12.0–14.0 **C. High** 0.80–1.10	4.6	Abnormal	Two days, continue same medication for 30 days	Physician advises
3 & 4 Alternate days	34.8	**High** 12.0–14.0	3.33	Normal	Continue same medication for 30 days	Physician advises
3,3 Alternate days	26.3	**High** 12.0–14.0	2.42	Low	Dose increased to 3–4 for one month	Physician advises
5,4 Alternate days	47.5	**High** 12.0–14.0	5.23	Abnormal	Two days, continue same medication for 30 days	Physician advises
5,5 Alternate days	52.4	**High** 12.0–14.0	6	Abnormal	Three days, continue same medication for 30 days	Repeat INR test after two days
2,2 Alternate days	22.1	**High** 12.0–14.0	1.6	Very low	IV with heparin injection to regulate blood circulation	Dalteparin (Fragmin®), Enoxaparin (Lovenox®), and Tinzaparin (Innohep®)

troubles continue to present difficulties to fruitful drug compromise in clinics. The Center for the Study of Health System Change (CSHSC) is a non-benefit association (HSC). More than 33% of the clinics reviewed utilized a mixture paper-electronic compromise method, attributable to disappointment with early forms of drug compromise capacities given by EHR suppliers. The prescription compromise was coordinated with automated affirmation and release requesting at emergency clinics with further developed EHR-based medicine compromise usefulness to build intelligibility, diminish information re-emergence, and advance more persistent well-disposed release instructions. Despite this, various issues remain, including further developing admittance to dependable medication chronicles, upgrading EHR convenience, better captivating doctors, and consistently trading patient data with the following suppliers of care. It will be fundamental to develop ways for critical stakeholders.

Further, patient security advocates, policymakers, specialists, it is EHR bot HER sellers, clinics, and clinicians share the best EHR plans and clinic

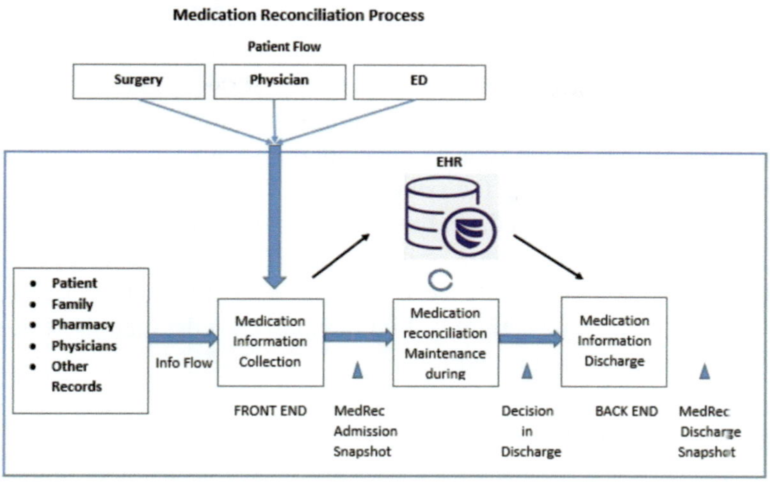

Figure 7.5 Medication reconciliation process.

execution methodologies to understand the possible wellbeing and proficiency advantages of EHR-based medicine compromise. They are separating data from the cycle log utilizing process mining strategies that assist with finding, screening, and working on the open course of prescription compromise. A back examination gives an organized cycle portrayal that can be applied to work processes to control the information extraction process, which helps in actual execution. The interaction mining strategy incorporates exploratory information investigation (EDA) and machine learning (ML) calculations. These are utilized to assemble information, process information, and ordering and information examination to further develop the cycle models [20].

Exploratory data analysis is the most popular method for doing first analyses on data to find patterns, identify anomalies, test hypotheses, and check hypotheses using outline measurements and graphical representations. The initial stage in examining and pre-processing is differentiating evidence of factors and information types. Non-Graphical Univariate Analysis is a form of univariate analysis that does not utilize diagrams to analyze essential measurements. Conversely, Univariate Graphical Analysis, often referred to as Bivariate Analysis, examines two elements: transformable factors and the management of exceptions, including their loss and treatment. AI (ML) is the logical investigation of calculations and factual models that PC frameworks use to play out a particular assignment without utilizing express guidelines, depending on examples and induction, all things equal. Integration with an electronic health record system could provide a complete picture of each patient's current medicine (including prescription and over-the-counter). The technique used in this prototype makes it applicable to any nation or location. The cost/utility of the system will be better than the labor costs of present medical specialists carrying out the same activity. The scalable improvement of patient safety, delivery speed, and quality may be achieved by the automated collection and reconciliation of drug data. This prototype can be used in any nation or area, and it will exceed current medical experts in terms of cost and utility for the same activities. Figure 7.5 gives a detailed illustration of the medication reconciliation process.

7.9 Classification Model

The focus is on building a classification model to predict adverse drug events. Data mining provides accurate predictions by considering the massive volume of data not on hypothesis testing but instead on repeated patterns of values of future outcomes. The machine learning algorithm can detect relationships between predictors and outcome variables to find accurate predictions [21]. The forecast is the classification model process that investigates the attribute's existing and previous states and its upcoming state. Classification is a kind of predictive model that can be used to predict repeated patterns of values to make accurate predictions and forecast future unlabeled records. We have used a decision tree induction algorithm to predict adverse drug events. There are two phases in decision tree algorithm: (1) training phase – a model is built from training instances; (2) testing phase – allocate a label to an unlabeled test instance [22, 23, 24].

7.10 Decision Tree Algorithm

The decision tree algorithm is used in the strategies for administering AI. It is frequently used to address an order problem or a relapse problem. The goal of this approach is to develop a model that can forecast how much an objective variable will be valued. The choice tree solves this problem by using a tree representation, where the leaf hub corresponds to a class name and attributes are discussed on the inner hub of the tree. The decision tree approach is used in the class that administers learning. The issue is resolved by employing a tree depiction, which compares each leaf hub to a class grade and addresses the attributes of the trees inside each hub.

Entropy: In a variety of situations, entropy is a measure of impurity, disorder, or uncertainty.

Entropy's purpose:

Entropy's formula is as follows:

$H(A)= -$probability of $\log_2(pc+)$ $-$probability of $\log_2(pm-)$,

where $(pc+)$ is the percent of positive class (yes) and $(pm-)$ is the percent of negative class (no).

In this scenario, the pure subset is a situation in which we will receive either all yes or all no answers. We have done this for one node, but what if, after splitting F2, we need some other attribute to get to the leaf node, and we need to add up the entropy of all those values to make the submission of all those entropy values? That is where the concept of information gain comes in.

Obtaining information: Information gathered is used at each stage of the tree-building process to decide which feature to split on. Since simplicity is preferred, we want to keep our tree as small as possible. At each phase, we should select the split that results in the purest daughter nodes. Each tree node's information value describes the level of detail that a feature offers regarding the class. The split with the greatest information gain will be the first one, and the process will continue until either all of the kid nodes are pure or the information gain is zero. The following steps are used for information gain formula: (S1, S2, S3, S4, S5, S6, and S7 represent steps)

S1: Determine the tree's root.
S2: Work out the entropy for each of the classes.
S3: For each attribute, calculate the entropy after splitting.
S4: For each split, calculate the information gain.
S5: Perform the splits.
S6: Increase the number of splits.
S7: Bring the decision tree to a close.

Algorithm to predict an adverse drug
Define Decision (Normal, Increase, Decrease, Mechanism, Suggested Management) Input: Parameters of Input data Output: if INR<=3.5:   ***if Drug=='warfarin 3mg':***   ***if adverse drug effect =='no effect'***   ***return 'normal'***

```
elif drug=='coagulant drug':
if adverse drug effect ==' breathing problem' return
'INR low'
elif drug=='coagulant drug':
if adverse drug effect=='abdominal pain'
return 'low'
elif INR>=5: if adverse drug effect=='Bleeding, Nau-
sea, Vomiting
 return 'very high'
```

7.11 Experiments and Results

The increase in warfarin dose will also increase or decrease INR levels, which indicates coagulation of the drug. With these techniques, predicting adverse drugs is possible. The following unwanted medicines have been predicted: Acarbose, Acetaminophen, Allopurinol, Amprenavir, ASA10, 15, Atazanavir, Azathioprine and mercaptopurine, Etravirine, concore, bisoprolol, and metoprolol. These techniques will give an automated system to detect adverse medication in health records so that physicians can easily prescribe the exact medication for the patient. INR level analysis with warfarin drug is presented in Figure 7.6.

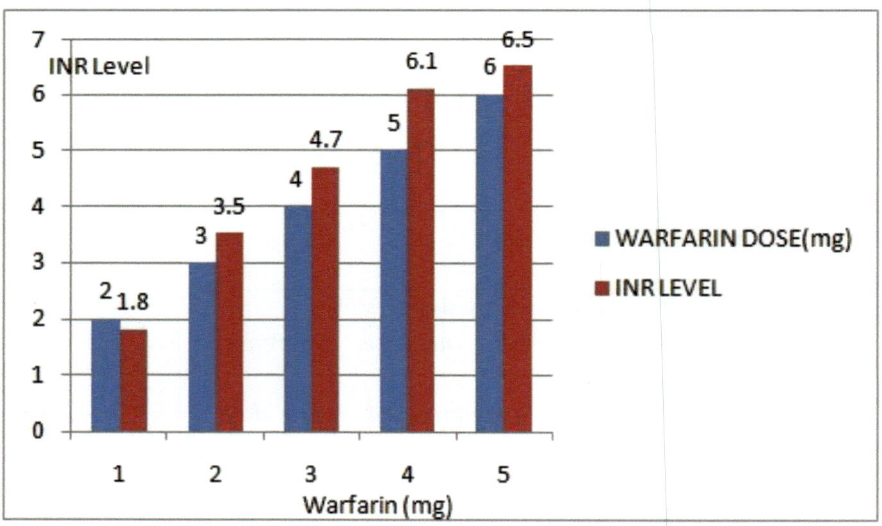

Figure 7.6 Analysis of increasing the INR level with warfarin medication.

7.12 Conclusion

Blockchain-based EHR computerized physician order entry systems includes a medication reconciliation process that will reduce the medication error and adverse drug events in EHR. We presented in this chapter the calculation of the risk ratio with confidence interval and performed subgroup analysis for categorical factors such as type of system, the system's functionality, study design, patient group, comparison group, level of care, and detecting error method. Create an accurate best possible medication history that compares existing medication orders against admission, discharge, and documented medication orders, resulting in the medication reconciliation process in a data mining method to provide patient safety. Here we conclude that blockchain-based EHR medication reconciliation process can offer an automated system to find adverse medication in the health record, which can reduce the risk of medication error and adverse drug events so that the physician can easily prescribe accurate medication for the patient, which will provide the patient safety in healthcare.

References

[1] Muthukumaran, V., Joseph, R. B., & Uday, A. K. (2021). Intelligent Medical Data Analytics Using Classifiers and Clusters in Machine Learning. In Handbook of Research on Innovations and Applications of AI, IoT, and Cognitive Technologies (pp. 321-335). IGI Global.

[2] S.Satheeshkumar, Dr.ManjulaSanjaykoti, Improved Quality of Patient Care and Data Security using Cloud Crypto System in EHR, ISSN 2460 4010 (ICCS) (2018), [Pg-169].

[3] Emmanuel KusiAchampong, Electronic Health Record (EHR) and Cloud Security: The Current Issues, International Journal of Cloud Computing and Services Science (IJ-CLOSER) Vol.2, No.6, (2013), [Pg. 417-420]

[4] Muthukumaran, V., Arun, M., Kumar, S. S., Kumta, S. D., Kavitha, M. A., & Vijayaraghavan, R. (2021, July).

[5] Secure efficient signature for internet of things over near-ring. In Journal of Physics: Conference Series (Vol. 1964, No. 2, p. 022015). IOP Publishing.

[6] Balavinothini, Ggnanambigai, A review on text mining in cloud using feature selection techniques, International Journal of Advances in Electronics and Computer Science, Volume-3, Issue-6, (2016) [pg4-6].

[7] Manikandan, G., Perumal, R., & Muthukumaran, V. (2021). Secure data sharing based on proxy re-encryption for internet of vehicles using seminearring. Journal of Computational and Theoretical Nanoscience, 18(1-2), 516-521.

[8] DeevakarRogith, Process Mining of Medication Revisions in Electronic Health Records University of Texas Health Science Center at Houston, UT SBMI Dissertations (Open Access) School of Biomedicao Informatics, (2015) [pg-5-9].

[9] Kumar, V. V., Muthukumaran, V., Ashwini, N., Beschi, I. S., Gunasekaran, K., & Niveditha, V. R. (2022). An Efficient Signcryption Scheme Using Near-Ring Hybrid Approach for an IoT-Based System. International Journal of e-Collaboration (IJeC), 18(1), 1-31.

[10] Kumar, S. S., & Koti, M. S. (2021). "An hybrid security framework using internet of things for healthcare system. Network Modeling Analysis in Health Informatics and Bioinformatics, 10(1), 1-10.

[11] Nick Gehrke, Nordakademie, chai for information system, Process Mining analysis, conformance checking, compliance checking, (2015)

[12] AshkanSharabiani, AdamBress, ElnazDouzali,andHoushangDarabi, "Revisiting Warfarin Dosing Using Machine Learning Techniques", Computational and Mathematical Methods in Medicine, Article ID 560108, (2015), [pg- 9].

[13] Raghunath, K. K., Kumar, V. V., Venkatesan, M., Singh, K. K., Mahesh, T. R., & Singh, A. (2022). XGBoost Regression Classifier (XRC) Model for Cyber Attack Detection and Classification Using Inception V4. Journal of Web Engineering, 1295-1322.

[14] R.S Mans W.M.P van der Aalst, R.J.B vanners., A.J Moleman, department of Infirmation system, Eindhoven university of technology, Process mining in healthcare data changes when answering frequently posed questions, (2012), [Pg-108].

[15] Karthick Raghunath, K. M., Koti, M. S., Sivakami, R., Vinoth Kumar, V., NagaJyothi, G., & Muthukumaran, V. (2022). Utilization of IoT-assisted computational strategies in wireless sensor networks for smart infrastructure management. International Journal of System Assurance Engineering and Management, 1-7.

[16] Muthukumaran, V., Vinothkumar, V., Joseph, R. B., Munirathanam, M., & Jeyakumar, B. (2021). Improving network security based on trust-aware routing protocols using long short-term memory-queuing segment-routing algorithms. International Journal of Information Technology Project Management (IJITPM), 12(4), 47-60.

[17] FadouaKhennoua, YounessIdrissiKhamlichib, Improving the Use of Big Data Analytics within Electronic Health Records: A Case Study based OpenEHR, The First International Conference On Intelligent Computing in Data Sciences, ScienceDirect, Procedia Computer Science, volume127,(2018), [pg60–68].

[18] Sadhasivam, J., Cera, J. M., Deepa, R., Satheshkumar, K., Muthukumaran, V., Kumar, S. S., & Kavitha, M. A. (2021, July). Movie recommendation system using clustering mining with Python. In Journal of Physics: Conference Series (Vol. 1964, No. 4, p. 042073). IOP Publishing.

[19] Kumar, V., Niveditha, V. R., Muthukumaran, V., Kumar, S. S., Kumta, S. D., & Murugesan, R. (2021). A Quantum Technology-Based LiFi Security Using Quantum Key Distribution. In Handbook of Research on Innovations and Applications of AI, IoT, and Cognitive Technologies (pp. 104-116). IGI Global.

[20] DeevakarRogith, Process Mining of Medication Revisions in Electronic Health Records University of Texas Health Science Center at Houston, UT SBMI Dissertations (Open Access) School of Biomedicao Informatics, (2015), [pg-5-9].

[21] Reddy, K. H. K., Luhach, A. K., Kumar, V. V., Pratihar, S., & Roy, D. S. (2022). Towards energy efficient Smart city services: A software defined resource management scheme for data centers. Sustainable Computing: Informatics and Systems, 100776.

[22] Kumar, V. V., Raghunath, K. M., Muthukumaran, V., Joseph, R. B., Beschi, I. S., & Uday, A. K. (2021). Aspect based sentiment analysis and smart classification in uncertain feedback pool. International Journal of System Assurance Engineering and Management, 1-11.

[23] Ahmed, S. T., Kumar, V. V., Singh, K. K., Singh, A., Muthukumaran, V., & Gupta, D. (2022). 6G enabled federated learning for secure IoMT resource recommendation and propagation analysis. Computers and Electrical Engineering, 102, 108210.

[24] Bell, Chaim University of Toronto, St Michaels, Hospital Medication reconciliation in Acute care, (2011), [p-15].

[25] AshkanSharabiani, AdamBress, ElnazDouzali,andHoushangDarabi, "Revisiting Warfarin Dosing Using Machine Learning Techniques", Computational and Mathematical Methods in Medicine, Volume 9, Article ID 5601082015.

[26] Sadhasivam, J., Muthukumaran, V., Raja, J. T., Joseph, R. B., Munirathanam, M., & Balajee, J. M. (2021, July). Diabetes disease prediction using decision tree for feature selection. In Journal of Physics: Conference Series (Vol. 1964, No. 6, p. 062116). IOP Publishing.

[27] Kumar, V. V., Raghunath, K. K., Rajesh, N., Venkatesan, M., Joseph, R. B., & Thillaiarasu, N. (2021). Paddy Plant Disease Recognition, Risk Analysis, and Classification Using Deep Convolution Neuro-Fuzzy Network. Journal of Mobile Multimedia, 325-348.

[28] Mahesh, T. R., Dhilip Kumar, V., Vinoth Kumar, V., Asghar, J., Geman, O., Arulkumaran, G., & Arun, N. (2022). AdaBoost Ensemble Methods Using K-Fold Cross Validation for Survivability with the Early Detection of Heart Disease. Computational Intelligence and Neuroscience, 2022.

[29] Selvi, G. V., Muthukumaran, V., Kaladevi, A. C., Kumar, S. S., & Swapna, B. (2022). Integrated dominating and hit set-inspired unequal clustering-based data aggregation in wireless sensor networks. International Journal of Intelligent Computing and Cybernetics.

[30] Wei DaiWeiji, A MapReduce Implementation of C4.5 Decision Tree Algorithm A MapReduce Implementation of C4.5, International Journal of Database Theory and Application Vol.7, No.1, (2014), [Pg.49-60].

[31] Kumar, S. S., & Koti, M. S. (2021, December). Efficient Authentication for Securing Electronic Health Records using Algebraic Structure. In 2021 5th International Conference on Electrical, Electronics, Communication, Computer Technologies and Optimization Techniques (ICEECCOT) (pp. 366-370). IEEE.

8

Blockchain-Based IoT Healthcare Data Using Narrow Elliptical Curve with Particle Swarm Optimization (NECPSO) Approach

Sivanantham Kalimuthu[1], P. Blessington Praveen[1], J. Mahalakshmi[1], R. Mohan Kumar[2], and S. Balaguru[3]

[1]Crapersoft, India
[2]Sri Ramakrishna Engineering College, India
[3]JKK Nataraja Engineering College, India

Abstract

Blockchain technology was meant for securing the data transfer between two users or parties using a cryptographic action and in the presence of a distributed ledger. Conventionally, it is known that IoT deploys a centralized architecture for better control where the data propagation is carried out by the cloud system over multiple heterogeneous devices. There is also various analytics to understand and extract certain useful information from this data transaction. As there is a massive number of devices connected using the IoT network, adopting such a centralized controlling system will restrict the operation. Hence, scalability has to go low evidently. This causes the system to undergo various vulnerable scenarios, which leads to a security breach. Therefore, the blockchain network's smart mechanism helps permit this IoT device to operate in a highly secure manner. It also constructs various agreements that are implemented when required. Applying blockchain to IoT devices has various benefits that assist in higher scalability, cost-effectiveness, and automation. It also assists in resisting different forms of overriding options by accident. The entire IoT device's obtained information is transmitted to the different destination nodes in a highly decentralized manner over the encrypted communication channel. The narrow elliptical curve with particle swarm optimization algorithm (NECPSO) model enforces a new cryptographic-based service that converts user-defined content to an incomprehensible format. This application model focuses mostly on the development of computationally secure key generation. As the word implies, the key used in the cryptographic process should meet two critical criteria: the cost of breaking the cipher data should be greater than the value of the information encrypted, and the time required to break the cipher should be greater than the information's full life span. The primary goal of this application model is to enable computationally safe key creation in order to encrypt data. This research effort divides key generation in the cryptographic process into three types. In the first section, a NECPSO encryption service model is built, in which the key generation is based on the traditional encryption operation mode with certain time and data size enhancements. When compared to existing services, comparative analysis demonstrates the efficiency of the proposed pioneering application model's quality and strength.

Keywords: Narrow Elliptical Curve, Particle Swarm Optimization, Healthcare IoT Data, Encryption and Decryption.

8.1 Introduction

IoT is a sophisticated form of networking of different computing and communication devices using unique identifiers of each device. To achieve a collaborative environment, IoT integrates various technologies such as hand-held digital devices, industrial machinery, animals (or) people, or any physical-digital entities to work together to achieve the application's purpose. However, there are various security issues including denial-of-service (DoS) attacks, data theft (or) data breaches, botnet attacks where several systems to take control of the victim's system, to extract the victim's confidential data this consistently matter of concern within an IoT (Kettani Houssain, and Polly Wainwright., 2019). Unsecured devices are the main security concern in frameworks of edge computing and IoT as they are used to increase network coverage. There are also devices within an IoT that can migrate from one network to another, which causes a serious threat to security. In present times, the conventional IoT network considers the concept of zero trusts, which is a mechanism for safeguarding the infrastructure of the Internet and the IT system. It will also mean that it allows any form of devices to join the network or access the resources within a network that assumes that they are also authorized. Hence, it is not difficult for the adversaries to go parallel with the regular node after bypassing the firewall system in an IoT. This is potentially a challenging situation as there are various IoT devices, and most of them are highly vulnerable and unsecured. This phenomenon makes an adversary job easier to utilize the gateway system present in an IoT environment (Tehrani et al., 2014).

At present, there are various taxonomies of the layer-based IoT operation, viz. (i) three-layered architecture, (ii) four-layered architecture, and (iii) five-layered architecture. The three-layered architecture consists of the application layer, network layer, and perception layer, while the four-layered architecture has the application layer, support layer, network layer, and perception layer. Five-layered architecture consists of the business layer, application layer, processing layer, transport layer, and perception layer. A closer look at all the taxonomies shows that three layers are the most frequently adopted layer in most IoT applications. Therefore, the proposed system considers choosing a three-layered architecture (Nawir, Mukrimah et al., 2016).

Figure 8.1 highlights the three-layer architecture where the application layer is used to define all the applications deployed in IoT and is generally inflicted with various intrusions, e.g., malicious codes and cross-scripting attacks. The blockchain usage is specific to a particular layer when it comes to positioning it in IoT layers. The three layers are perception layer network layer, and application layer. All the stream of the request arrives from the different nodes in the perception layer, which is then forwarded to the network layer where the decentralized mechanism of operation of blockchain resides in the decentralized cloud. The decentralized cloud further communicates with the cloud with respect to a different application for administration over the application layer. The blockchain application is hosted between the network and application layers, whose main responsibility is to organize communication between communicating nodes for IoT scenarios (Viriya-sitavat Wattana et al., 2019). The centralized server of the cloud will be responsible for further processing of the data obtained. Further, the SDN switches reside over the perception layer where the stream, as well as request flow, are assessed, and based on this operation, the information flow is passed over the controller, which resides in the network layer, and the controller communicates further with the cloud server to run an algorithm that justifies the regular or malicious nature of the communication link. The result of this communication yields a secured IoT with SDN-based communication system secured by blockchain operation. However, it depends upon the sensitivity of the application.

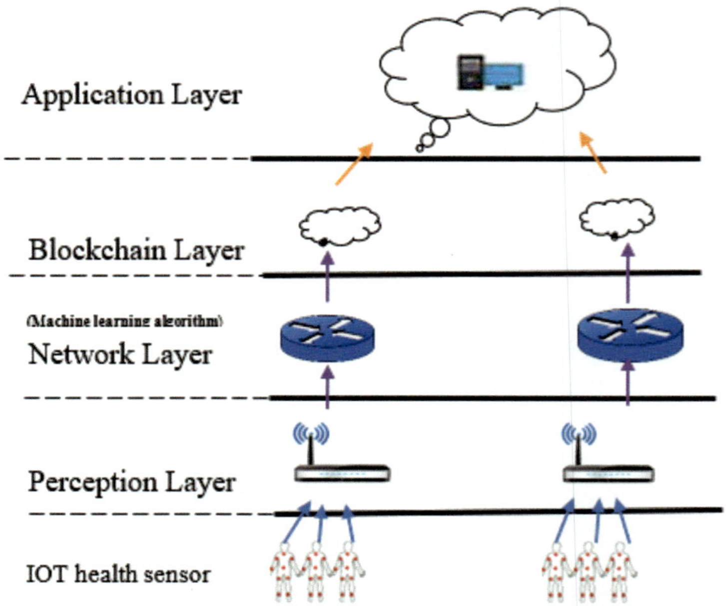

Figure 8.1 Position of the blockchain in IoT layers.

The network layer is used for transmission purposes while forwarding all the data obtained from the IoT device. The majority of the communication protocol runs in this layer. The intrusion in such layers is much, viz. denial of service attack, man-in-middle attack, storage attack, and exploit attack. The perception layer is basically the bottom layer, and it is responsible for extracting the raw sensory data and pushing it forward to the network layer. Generally, this layer is susceptible to eavesdropping replay attacks and node capture attacks (Xie Lixia et al., 2019). The proposed study focuses on developing a secure communication scheme where blockchain plays a significant role; therefore, the study uses the network layer as the host to reside in the three algorithmic operations. Other layers (i.e., application and perception) can be automatically secured if the network layer acting as an intermediate bridge is highly secured.

The next security issue in an IoT environment is its low visibility of the network. At present, the utilization of the devices in an IoT is increasing exponentially, that is, it is quite a difficult task to offer generalized security for all the devices. The conventional security practices in information technology call for accounting for all the components connected to the network; however, it is not feasible in IoT due to many IoT devices and mobility factors. The relationship between security and mobility is yet to be solved (Bandyopadhyay Debasis and Jaydip Sen, 2011). Figure 8.2 highlights the different forms of IoT security issues. The next security problem is associated with the software's vulnerabilities in an IoT. For example, the telecommunication service providers (TSPs) are wide-open to cyber-attacks implemented by bargained IoT sensors associated with their user networks. Such cyber-attacks will influence the target systems' intense effects and the telecommunication service providers themselves (Sivanantham Kalimuthu, 2021). The combined usage of different protocols and different IoT devices also calls for using four different overlapping software. Each software is characterized by a unique feature and vulnerabilities that can be misused by an adversary. It will also result in downtime due to the generation of errors when interacting with each other or resulting in more vulnerable data. Although updating the software could significantly assist in thwarting the attacker, updating such massive software is not possible concurrently. It will not only induce downtime but also renders the complete network vulnerable.

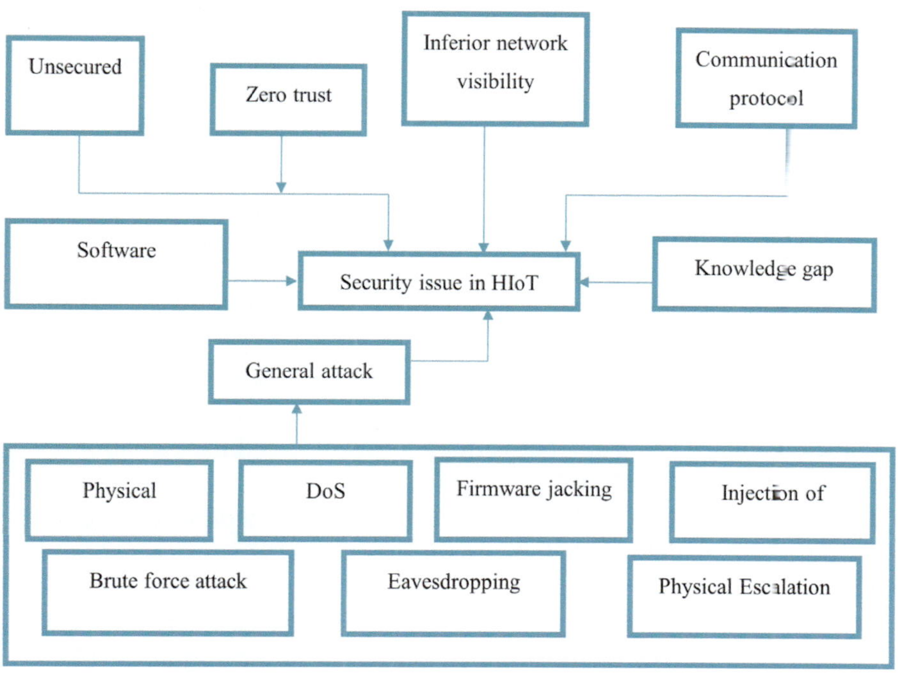

Figure 8.2 Taxonomy of security issues in a healthcare_IoT (HIoT).

Adopting open-source software is another frequently adopted step in many organizations to essential resources; however, such open-source software is quite easy to be compromised by the attacker (Zhang Zhi-Kai et al., 2014).

Another essential security issue in the IoT environment is a knowledge gap among the user. The users cannot be anticipated to know all the massive forms of attack strategies and vulnerable environment. The users are more concerned with the convenience of using IoT devices, and they are least bothered with the security flaws associated with the usage of such devices. The communication channel is another reason for security concerns in IoT (Kalimuthu Sivanantham et al., 2021). There are higher possibilities of originating attacks from the communication channel utilized by an IoT component connected variously to each other. At present, various protocols are used in an IoT communication system that already has a reported security issue and is known to negatively influence the complete networking system (Vimalachandran Pasupathy et al., 2020). There are various forms of attacks in an IoT; however, some of the common attacks in an IoT environment are:

- **Physical tampering:** Threats to unauthorized tampering on the physical device are possible when the devices run in an environment that cannot be accessed by the people.
- **Denial-of-service:** It is one of the most reported attacks in both cloud environment and IoT where the complete network and servers are paralyzed, rendering it easier for attackers to access the network or fulfill other malicious intentions.
- **Firmware jacking:** If the network is compromised, then the updates for upgrading the firmware are also possibly compromised, and when such compromised firmware update reaches hardware, it infects them too as there are failed authentication of the source of firmware update origination point.
- **Injection of malicious node:** The malicious nodes are physically deployed by the adversary among the regular normal IoT devices. The malicious nodes can carry out various forms of illegitimate control of the network and illegal sniffing of data.
- **Brute-force password attack:** Such attacks are also deployed to access the devices in vulnerable IoT environments.
- **Eavesdropping:** The vulnerable communication channel can be compromised and intercepted by the adversary node. The adversary can steal all the sensitive data from a certain weak spot in the communication channel.

- **Physical escalation:** Various forms of the flaws associated with the operating platforms, hardware, and software can be misutilized by the attacker to have illegitimate access to the IoT resources.

Therefore, the existing communication and deployment environment in an IoT is covered with various security issues (Abomhara Mohamed and Geir M. Køien, 2015). Various attacks reduce the trust factor of using the application over an IoT from a user perspective. The next section will provide background briefs on existing security approaches. At present, blockchain technology is seen as the long-term security approach in an IoT because of multiple reasons. Blockchain induces information bound in the form of a block and are connected using cryptography. A chain of blocks is made where the Secure Hash Algorithm (SHA-512) hashed value of preliminary data (or block) is retained in one block along with transaction data in tree form and timestamp. Blockchain is characterized by stability, traceability, process integrity, security, and faster processing. However, there are variously reported limitations to blockchain, i.e., higher energy consumption, immutable data, self-maintenance dependency, higher cost, and the concept are still in the nascent stage of development. However, this limitation can be overcome if there is enough focus on the encryption mechanism used in liking blocks (Sivanantham Kalimuthu et al., 2022). The proposed system uses the NEC cryptosystem with particle swarm optimization for this purpose, an asymmetric form of encryption. Owing to the faster mode of encryption and decryption process with a highly reduced number of steps, the proposed system uses the NEC encryption approach's potential to link the blockchain. The remainder of this book chapter is organized as follows. Section 8.2 discusses past blockchaining and IoT for various datasets and related work, Section 8.3 discusses the proposed narrow elliptical curve with particle swarm optimization (NECPSO), and Section 8.4 compares the experimental outcomes of proposed NECPSO systems. Finally, Section 8.5 contains the work's concluding thoughts and future scope.

8.2 Literature Review

At present, the IoT security solutions are highly specific to the different layers of the IoT protocol stack explained in Figure 8.3. The application layer in an IoT offered constrained application protocol (CoAP) and secured message queue telemetry transport (MQTT), which uses an encryption mechanism over the traffic data. The transport layer mainly offers TCP and UDP and

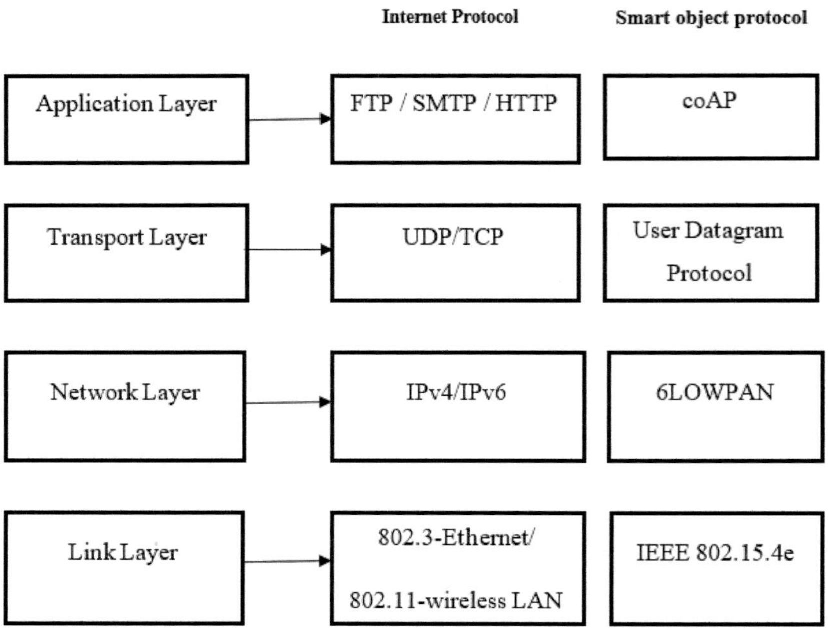

Figure 8.3 Protocol suites for smart objects in IoT.

DTLS and SSL to offer security over the transport layer. The network layer uses IPv6, 6LoWPAN protocol to secure any data and is mainly meant for transmission. The link-layer uses the IEEE 802.15.4e standard, which has its default mechanism of traffic management (Aleshi Ahrash et al., 2019).

There are various research works, e.g., where security approaches have been discussed for IoT. Existing approaches toward securing IoT mainly use an authentication-based approach, key management, privacy preservation, fault tolerance-based approach, and an explicit mechanism for resisting specific attack. Apart from this, performing sophisticated security operations over devices with low-end processing capability is not practical. Another prominent security issues in IoT is the little emphasis on security standards. A closer look into the existing security solution shows that they are more inclined toward strengthening privacy and less to data integrity, service availability, maintainability, etc. Another bigger challenge is the usage of the encryption approach. Usage of the encryption approach is mandatory for better security, but they are not optimally used, and their strength is restricted in IoT in the presence of dynamic attackers (Roman et al., 2013).

This section deals with the existing research works in the domain of IoT with the encryption-based security approaches. The work of Awan Ijaz Ahmad et al. (2020) described a secure communication mechanism by improving the conventional advanced encryption standard (AES) algorithm. The encryption mechanism resists different security attacks and also minimizes the encryption power. The system has not addressed the issues of information-centric IoT. Sufficient data protection can be achieved using attribute-based encryption (ABE) mechanism consisting of cipher text policy introduced in Chen Cheng et al. (2011). The encryption mechanism gives improved performance and minimized computation overhead. The resource-constrained IoT devices are composed of different issues like low computational ability, high latency, low bandwidth, etc. These issues are addressed in Mahmud Redowan et al. (2018) with a lightweight encryption technique based on cellular automata for IoT. The technique can reduce the run time and can prevent data theft. The complexity parameter is not addressed in prior work. The work of Rajesh Sreeja et al. (2019) has presented a reduced block encryption mechanism for the IoT where less complex and fast encryption is achieved. The block encryption mechanism yields reduced computations and offer higher security that supports low-powered IoT devices. The work of Yang Yuhan et al. (2021) has introduced a multiparty authentication-based encryption mechanism for narrowband IoT terminals. The encryption mechanism is based on certificates policy, which provides data privacy and efficient support for the 5G network. The chaotic encryption algorithm is introduced in Hu Bin et al. (2018) for intelligent data transmission in IoT with key controlled neural networks. The cryptanalysis gives improved information entropy than the existing algorithm. The summary of the above research works is given in Table 8.1.

The following section describes the various research works toward blockchain-based IoT security. The smart applications of the IoT face many security as well as access control challenges. Hence, Wang Jian et al. (2021) proposed a blockchain concept with a lightweight proof-of-block and trade approach for large-scale IoT systems. This security analysis approach yields significant improvement in the system performance by offering security with reduced computational time. The blockchain can help perform IoT data access but fails to handle large IoT data due to less storage space. Hence, Shi Peichang et al. (2019) proposed a lightweight architecture for blockchain. This architecture's outcome gives reduced overhead in storage cost under both stochastic and worst-case scenarios. The blockchain can offer proper

Table 8.1 Summary of study toward encryption-based security approaches for IoT.

Authors	Problems	Technique	Advantage	Limitation
Awan Ijaz Ahmad et al., 2020	Security of IoT	AES algorithm	Resist different security attack	Not addressing the issues of information-centric IoT
Chen Cheng et al., 2011	Sufficient data protection	ABE algorithm	Improved performance and minimized computation overhead	Not considered resource-constrained IoT device
Mahmud Redowan et al., 2018	Resource-constrained IoT devices	Lightweight encryption technique	Reduce the run time and prevents data theft	Complexity parameter is not addressed
Rajesh Sreeja et al., 2019	Data encryption	Reduced block encryption	Less complex and fast encryption	Not benchmarked
Yang Yuhan et al., 2021	Authentication	Multiparty authentication based encryption	Data privacy end effective support for 5G network	A comparative analysis is not conducted
Hu Bin et al., 2018	Intelligent data transmission	Chaotic encryption algorithm	Improved information entropy	Application aspects are not discussed

privacy and security solutions with significant computing but fails to integrate IoT computing scalability. Hence, Biswas Sujit et al.'s (2018) scalable group chain concept is introduced for IoT services' blockchain structure. The group chain mechanism helps to achieve better transaction output through optimization and reduced latency (Sivanantham, Kalimuthu, 2022). Adil Muhammad et al. (2021) have addressed the IoT authentication issue with a multi-WSN authentication scheme based on the blockchain. The outcome of this authentication scheme gives a better security solution and performance. Blockchain offers high security and creditability; hence, Zheng Bao-Kun et al. (2018) have introduced a threshold IoT service based on blockchain. The outcomes of it show reduced response time and yield better security. Large-scale networks face various security issues in IoT that need to be addressed with a proper security solution. Wei Jianghong et al. (2016) have presented an attribute-based access control mechanism that resists multiple attacks and offers better security. The summarized form of the above research papers is tabulated in Table 8.2.

Table 8.2 Summary of study toward blockchain-based IoT security.

Authors	**Problem**	**Technique**	**Advantage**	**Limitation**
Wang Jian et al., 2021	Access control challenges	Lightweight proof of block and trade approach	Security with reduced computational time	Not performed a comparative analysis
Shi Peichang et al., 2019	Handling larger IoT data	Lightweight architecture for blockchain	Reduce overhead in storage cost	Not involved in benchmarking
Biswas Sujit et al., 2018	IoT computing scalability	Scalable group chain concept	Achieve better transaction output through optimization and reduced latency	Not performed a comparative analysis
Adil Muhammad et al., 2021	IoT authentication issue	Multi-WSN authentication scheme	Better security solution and performance	Not benchmarked
Zheng Bao-Kun et al., 2018	Security and creditability	Threshold IoT service based on blockchain	Reduced response time and yield better security	Not performed a comparative analysis
Wei Jianghong et al., 2016	Security issues in IoT	The attribute-based access control mechanism	Resist multiple attack and offer better security	Not benchmarked

Intelligent transportation is a form of vehicular IoT application, but it needs a proper reliable security solution against user privacy data attacks. Hence, Lin Jie et al. (2017) investigated the transportation system's challenges in the IoT environment. The author has brought the blockchain concept for the vehicular IoT solution and gives a better privacy preservation mechanism. The blockchain can be incorporated in the SDN to solve the forensic issues addressed in Duy, Phan The et al. (2019). In this, the author has introduced the blockchain-based distributed controller in forensic architecture and achieves performance improvement with reduced delay, increased throughput, improved security, and accuracy. A work of Rathee Geetanjali et al. (2021) has described the inclusion of blockchain for IoT data security and trusted communication. However, the blockchain faces a drop rate issue in the IoT system. The research problems of the proposed study are as follows:

- Existing security solutions do not offer equal emphasis to the maximum-security standards in an IoT. More work is focused on achieving privacy, mainly leaving other standards vulnerable. So the proposed study identifies this problem and incorporates other security standards, e.g., anonymity, confidentiality, availability, integrity, and privacy.
- Adopting existing encryption practices for secure communication is not practical. There is no evidence to prove if they offer sustainability over devices with limited resources and balancing communication demands and security. To address this issue, the proposed study adopts both analytical and experimental models to offer proof of concept where blockchain and NEC algorithm is used for data security.
- Owing to the lack of appropriate intelligence-based security services, existing approaches cannot construct precise decisions for offering reliable resistance against threats of unknown type in IoT. To address this problem, the proposed system includes an optimization algorithm that is capable of offering more information and undertakes better decisions toward routing the data from the sensor to the IoT cloud.
- There is no doubt that blockchain technologies offer impressive security. Still, there is no denying that it also has various pitfalls that are not addressed in any existing studies. This issue is addressed by revising the conventional blockchain design where the legacy problems are overcome.

A closer look at the research shows that existing IoT communication system has the availability of various security operations; however, they all suffer from certain issues. Apart from this, existing encryption-based algorithms are highly attacked and have non-supportability for reliable and cost-effective transmission. A better security system can only be reshaped to smart and intelligent if more analytical data could be facilitated to the model, which is currently not found much in the existing approach. Therefore, the statement of the proposed study's problem is "Existing IoT security solution cannot offer a higher degree of reliable resistivity using conventional blockchain as well as it also lacks computational cost-effectiveness" (Wolfert Sjaak et al., 2017).

8.3 System Design

The proposed research work's prime aim is to carry out a secured communication scheme considering a suitable communication environment in an IoT.

The previous model discussed in this chapter presents a security framework using blockchain and NEC cryptosystem, focusing on the security aspect. However, the model discussed in this chapter targets achieving optimized security performance considering the global communication scenario in IoT equipped by SDN.

The proposed architecture is highlighted in Figure 8.4, which shows that the complete idea of implementation is classified into identification and prevention from the adversary present in IoT. The idea is also to assess the security aspect of the data flow. The implementation of the model discussed in Chapter 4 has used blockchain to discuss how the data are converted into blocks, further protected by the NEC cryptosystem. However, it has

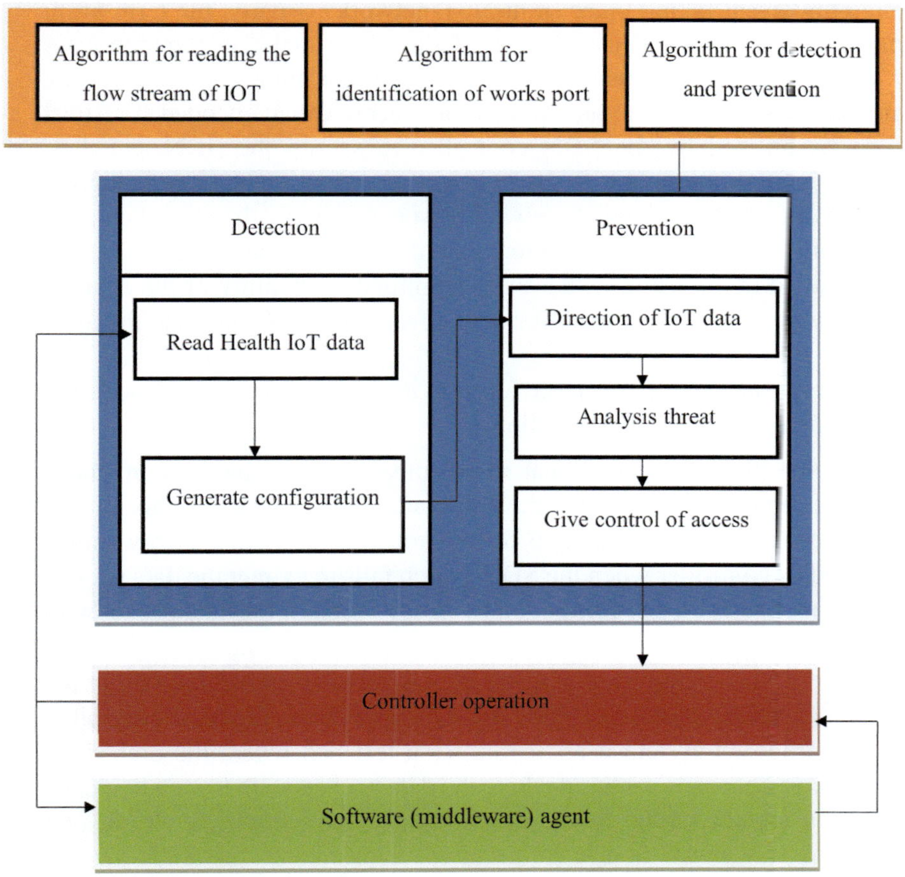

Figure 8.4 Architecture for optimized secure communication.

an inclusion of an encryption system, and the adoption of blockchain is further required to be improved from the computation. The proposed system addresses this research challenge by presenting an improved blockchain operation discussed in Chapter 4. At the same time, this chapter further offers an assessment of each link concerning the degree of vulnerability. Hence, this operation of assessing the vulnerability from a link perspective further optimizes the proposed system's security operation.

The proposed system uses the southbound protocol, which mainly connects switches within the data plane with the control plane. The study considers that the control link's message is mainly of two types, viz. (i) non-contemporary beacon and (ii) encoded statistical beacon. The SDN switch utilizes the non-contemporary beacon to intimate the SDN controller concerning any event taking place in the IoT network. Usually, the term event pertains to the arrival of any stream of data flow and the elimination of any specific data flow. This also results in a certain amount of cost for the complete SDN system. The second kind of beacon, called encoded statistical beacon, is utilized by the SDN controller to set up the SDN switches. The term configuration over the switch will essentially mean extracting the statistical information from the switch associated with the data flow and constructing up the entries of different data flows. One essential constraint to be understood in this point is that there is limited availability of the channel capacity while transmitting the information from the SDN control to the SDN switch.

The proposed system uses more forms of discrete beacons while modeling and used in algorithm implementation.

- The first type of beacon is called an embedded message beacon, and it comes under the category of the non-contemporary beacon. This kind of beacon (or control message) is formulated by the switch and forwards to the SDN controller when the SDN switch fails to extract the data packet. It cannot perform any form of equivalency check of the data packet with any existing information of the priory cached stream entries. All the data packets whose equivalency score is very low are embedded within the embedded message beacon and finally forwarded to the SDN controller. The proposed system uses a simulation test bed where the embedded message beacon is captured where the data packets with low equivalency scores are forwarded to the controller. At the same time, its header retains the source node's information connected to the SDN switch. In contrast, the complete data packets with low equivalency scores and memory information are retained within the embedded message beacon's payload.

- The second kind of beacon is termed a clean flow beacon, which also comes under the non-contemporary beacon category. This message is forwarded to the SDN controller by the SDN switch when the stream of data flow departs from the switch's egress port. The statistical information associated with the filtered data flow is forwarded to the SDN controller with the respective field's assistance by the proposed logic. The simulated environment is created in the proposed system to capture this message information. The first field of this message duration of data flow retains information about the duration of the filtered data flow. The second field of this message number of data packet retains information about the total data flow filtered out from the stream. The third field of this message cause-field represents the timeout's idle period, which justifies the filtering of stream entries due to lack of data packet within that particular data flow for a specific duration range.
- The third kind of beacon is termed an embedded statistical beacon (Figure 8.5), which is a part of the message when the SDN controller interacts with the SDN switch. This message is meant for performing encoding operation toward any form of request over switch or response, which has a massive quantity of data packet. It is not possible to forward them in one unit of data flow. SDN 6 controller mainly uses this embedded statistical beacon to obtain statistical information from the SDN switches. To obtain extensive information associated with the data plane (concerning switches), the SDN controller obtains the statistical information using this embedded statistical beacon from the SDN switches. A similar simulation testbed is constructed for this purpose. They can also be used to update operations over the various statistical information in the table associated with the data flow. Various information is maintained within each flow table, viz. equivalency attribute field, number of the data packet, and execution. The duration attribute is used to retain information about the operational time involved in the data flow. The number of data packet fields is used to retain information about the data packet with a high equivalency score concerning the data flow. The equivalency attribute field is meant for segregating and categorizing the types of data flow with better accuracy.

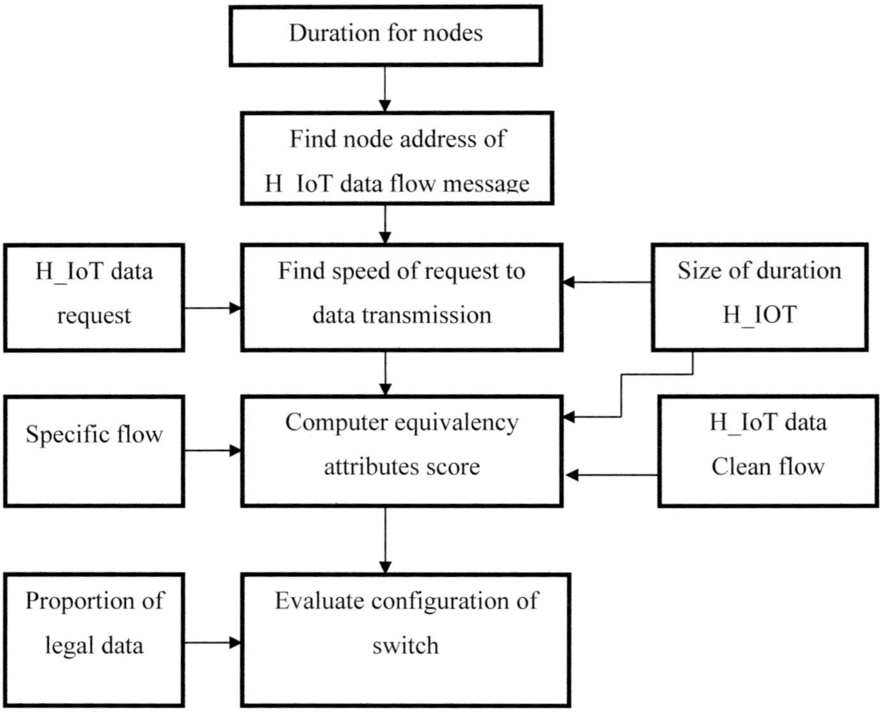

Figure 8.5 Process flow of reading the flow stream in H_IoT.

Using the messages mentioned above, the proposed system carries out a secure communication operation to resist any malicious data flow of unknown attack origin. For this purpose, a definitive adversarial model is developed, which carries out the introduction of various new data packets with low equivalency scores toward the SDN switch. The SDN switch considers such a new data packet with a low equivalency score as a new data flow by the working methodology of the SDN in the proposed IoT environment. Hence, such data packets are embedded within embedded message beacon and are forwarded to the SDN controller, which further performs the computation of the routing links and further transmits the interactive message (from the controller to SDN switch) to all the SDN switches associated with the routes for transmitting the new data flow and leads to configuring the new entries of stream information. The process of the proposed algorithm design (Figure 8.5) is explained in the following architecture diagram.

Narrow elliptical curve cryptosystem:

NEC cryptosystem is adopted for enhancing the process of randomization while performing the ciphering process using the mode of asymmetric encryption technique. The reputation of using this ciphering process is high in post-quantum cryptography. This algorithm's prime construction is based on linear codes and its decoding process connected with its complications. The proposed system chooses to use the NEC encryption system due to its essential features, making it different from other encryption approaches. As the proposed system is deployed over an IoT environment with many nodes, encryption must be lightweight. To be lightweight, an encryption algorithm should execute faster and occupy lesser memory. NEC algorithm facilitates faster ciphering as well as a deciphering process compared to the potentially strong algorithm AES. Although the NEC algorithm's legacy version does not offer any process to generate a signature, amending the conventional scheme could successfully result in a signature generation.

(i) Step-1 generation of key

This is the first step of the NEC algorithm. A selection of a linear code K_L is carried out by sensor node-A. This selection is carried out from the sensor node-A codes to be the best mechanism to perform decoding. It also renders the linear code L_c public while keeping the algorithm to perform decoding as private information. To carry out the decoding process, the proposed system has dependencies over two essential attributes, i.e., (i) information about linear code K_L and (ii) information about all the parameters that could be utilized while signifying linear code K_L from the selected group of codes. All this information is required to carry out the randomization process. The operation carried out in the process of generation of codes is as follows:

- A binary linear code K_L is selected as $(h1, h2)$, which can be potentially used for rectifying e errors obtained from the large group of codes. The system then evolves for generating a potential decoding process χ due to this selection process. Consider that F is a generator selected for the matrix of linear codes K_L while there are various generator matrices for all the linear codes K_L. However, there is a definitive selection associated with this group of codes as there is a possibility of disclosing the decoding process χ.
- An arbitrary value of dimension $(h1 \times h2)$ is selected by sensor node-A, a non-singular matrix of binary form B.

- An arbitrary matrix of dimension ($h1 \times h2$) is selected by sensor node-A, basically a matrix P form by permutation.
- The sensor node-A performs the computation of $h1 \times h2$ matrix $F1$, equivalent to BFP.
- The public key's formulation is done as ($F1$, t), and the private key is formed as BFP, where matrix P offers the possibility of encoding. It can be stored in the form of a specific parameter utilized for choosing linear key K_L.

(ii) Step-2: Encryption of H_IoT data

This is the second stage of the algorithm implementation, which considers that sensor node-B wants to forward a message *msg* to sensor node-A, a public key defined as ($F1$, t). Then the following operation takes place:

- A message *msg* is encoded by sensor node-B in the form of a string with the length of $b2$ of binary form.
- A vector L_{c1} is computed by sensor node-B as $K_{L1} = msg.F1$.
- An arbitrary message with the vector of size n-bit is generated by sensor node-B that possesses precisely t number of ones.
- An encrypted message is computed by sensor node-B as $K_L = K_{L1} + z$.

(iii) Step-3: Decryption of H_IoT data

This algorithm's decryption process is slightly different from other existing algorithms where the steps are not the same as the encryption process. Once the sensor node-A receives linear key K_L from sensor node-B, it performs the following operation:

- A computation of ($1/P$) is carried out by sensor node-A.
- An improved linear code is computed by sensor node-A as $K_{L1} = K_L$ ($1/P$).
- The decoding algorithm χ is used by sensor node-A for extracting L_{c1} to $msg1$.
- Finally, the decoded message is obtained by sensor node-B as $msg = msg1.(1/B)$.

Finally, the NEC algorithm also carries out proof of decryption of message to ensure that the message obtained by the end sensor node is having data integrity. For this purpose,

$$L_{c1} = L_c{\cdot}(1/P) = msg{\cdot}F1{\cdot}(1/P){+}z(1/P) = msg{\cdot}B{\cdot}F{+}z(1/P). \quad (8.1)$$

In this scenario, it is feasible for Goppa code F to rectify the t errors, and the message $msg.B.F$ is located at a distance of t from $K_L(1/P)$. Hence, the system can obtain rectified code as $msg1 = msg.B$. Therefore, if the inverse of B is now multiplied, then it will give

$$msg \;= msg1.(1/B) = msg.B.(1/B). \tag{8.2}$$

Therefore, the outcome is matching plain text, which proves that the decryption process is successful. Usually, the attacks that this algorithm can successfully resist are those adversaries with possession of the public key and not the private key. This algorithm can successfully resist brute force attack and structural attack. However, this algorithm has certain loopholes, mainly associated with the public key's size, which is a bit larger. It is seen that even after using liner key with the recommended parameters, the size of the public key will be around 312 bits, which is quite larger and will yield issues during implementation. Apart from this, the encrypted message is also much longer compared to the original message in plaintext. Usage of such a longer encrypted message will inevitably increase channel capacity consumption, resulting in a more vulnerable condition. This will also cause errors during transmission. Apart from this, the NEC cryptosystem is not applicable when it comes to the authentication process. It also lacks the signature scheme as the complete encryption process is not a one-to-one process, and the complete algorithm is highly asymmetric. It will mean that there is no way of achieving commutation between the process of encryption and decryption. Therefore, the proposed system revises this flaw using a unique security procedure discussed in the next chapter. The prime reason for this adoption is mainly its faster process of performing the ciphering and deciphering process. The following section explains the NEC with PSO hybridization and performance of blockchain encryption and decryption process.

Proposed NECPSO algorithm for blockchain encryption:

Encryption is followed by the process (steps 1–4)
1) File such as Executable, Document, DLL, AVI, and PDF from H_IoT are read as input files.
2) Split the Executable, Document, DLL, AVI, and PDF data individually.
3) Convert Executable, Document, DLL, AVI, and PDF data to binary bits.
4) Let us take the input as Xi and fill the block components Xi in 8 × 8 matrices till reaching the end of the file.

The key generation algorithm process is followed by (steps 5–14)

5) Generate the initialization vector using narrow elliptical curve, for the key generation process.

6) Use the upgraded cipher blockchaining process to make Yi = Xi * NEC.

7) The key of the successor block matrices is the resulting NEC matrix.

8) The pseudo random number generators generate yet another key matrix.

9) The most essential key is the next key generation process, which is built as a suboptimal key to acquire optimised encryption from the particle swarm optimization algorithm.

10) The key values are available in three different sizes such as 64, 128, and 256 bits. When the key size is 64, the block size is 8 × 8, when it is 256, the block size is 16 × 16, and when the key size is 128, the block size is 8 × 16.

11) The partially encrypted items in the matrices are again disjunct with this key value.

12) Completely ciphered data is available at the matrices in step 11.

13) The binary elements in the blocks are transformed to ASCII codes again.

14) The encrypted text data can be obtained by detaching the blocks.

Algorithm for decryption (steps 15–20)

15) As input, a file containing encrypted text data is used.

16) Convert the encrypted data to binary equivalents.

17) Fill the 8 × 8 matrices with the translated ciphered binary bits until you reach the end.

18) The process of encrypting in reverse, as well as key production, is known as reverse encryption.

19) Encryption is reversed as the process of decryption. Because the symmetric key technique is used, the decrypted, original text data is retrieved as a result of key generation.

8.4 Result and Discussion

The following experimental findings are obtained using the suggested hybrid encryption technique. The simulation environment of particle swarm optimization solver available in MATLAB R2013a and manual NEC algorithm is derived for the implementation. The encryption and decryption processes are carried out using Visual Studio 2010, C#, and a Windows 7 operating system with a Core-i3 processor and 3 GB RAM. 64-bit, 128-bit, and 256-bit keys are used. The following experimental results are achieved by using different key sizes. Based on the outcomes, a variety of matrices are examined. In the case of text encryption, time complexity is the most important issue to consider. Many forms of multimedia files are experimented and tested, and in the case of image encryption, various types of photographs are confirmed in this method. We conducted the following research to demonstrate the efficacy of the proposed picture encryption and its resistance to a variety of known assaults, including a known-plaintext attack, cipher text-only attack, statistical attack, differential attack, and various brute force attacks.

Any healthcare IoT data encryption computation should, as a minimum, defend against all types of referred assaults, such as the known-plaintext assault, the factual assault, the differential assault, the figure text-only assault, and various types of real-time attacks. For the inquiry, the proposed methods make use of various sizes of colored data. Visual examination, histogram examination, and correlation coefficient are the security investigation approaches used. All of these analyses show that the proposed procedures completely and appropriately satisfy all of the outcomes. The five types of data such as Executable, Document, DLL, AVI, and PDF are taken to the proposed algorithm cryptosystem. The consumed bytes of each file type is 12,228, 293,869, 991,961, 3,920,652, and 17,165,609 are explained in Figure 8.6 and Table 8.3.

The five types of data such as Executable, Document, DLL, AVI and PDF are taken to the proposed algorithm cryptosystem. Each file key size in bits is 512, 64, 256, 128, and 256 as explained in Figure 8.7 and Table 8.3. From the figure, exe file consists of the largest key size compared to other file types.

The five types of data such as Executable, Document, DLL, AVI, and PDF are taken to the proposed algorithm cryptosystem. The consumed encryption time in each file type in terms of milliseconds is 0.6, 0.148, 0.59, and 4.9

Table 8.3 Proposed NECPSO performance results in various file types.

File size (in bytes)	File type	Key size (bits)	Encryption time (in ms)	Decryption time (in ms)
12,228	Executable	512	0.66	0.61
293,869	Document	64	0.6	0.7
991,961	DLL	256	0.148	0.163
3,920,652	AVI	128	0.59	0.56
17,165,609	PDF	256	4.9	4.7

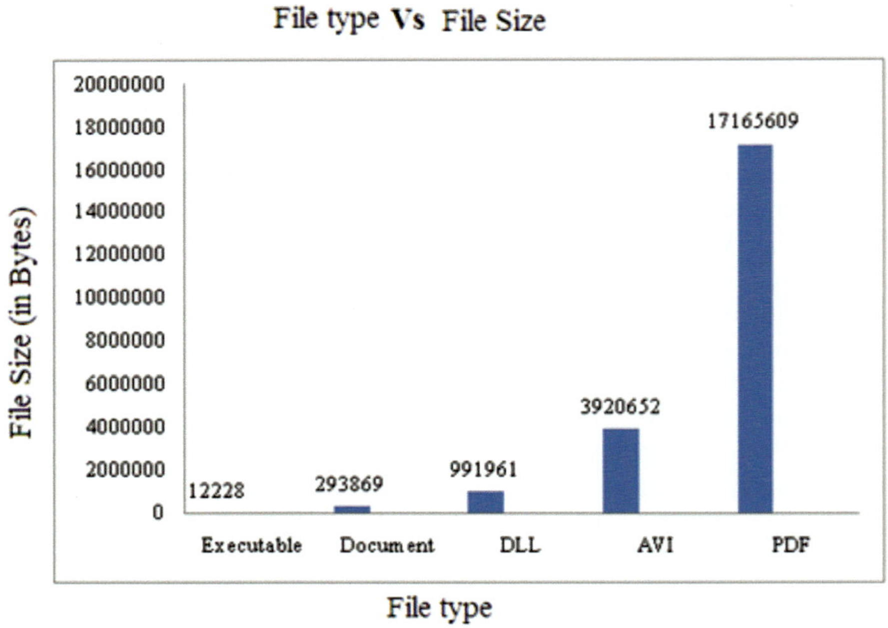

Figure 8.6 Comparisons for various file types vs. file size (in bytes) in NECPSO.

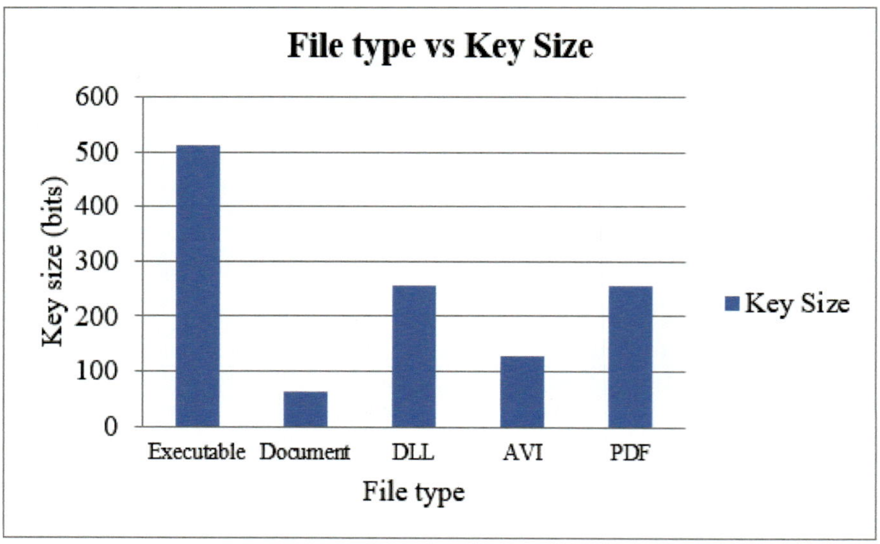

Figure 8.7 Comparisons for various file types vs. key size (in bits) in NECPSO.

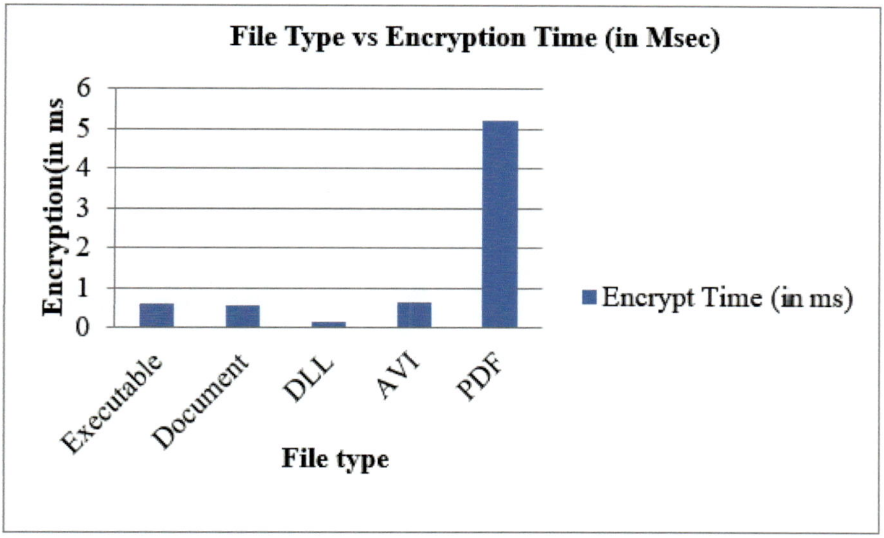

Figure 8.8 Comparisons for various file types vs. encryption time (in M.secs) in NECPSO.

as explained in Figure 8.8 and Table 8.3. From the figure, the PDF file encryption takes more time, i.e., 4.9 milliseconds, while other files consumed only below 1 milliseconds time duration.

The five types of data such as Executable, Document, DLL, AVI and PDF are taken to the proposed algorithm cryptosystem. The consumed decryption time (in M.secs) in each file type in terms of milliseconds is 0.61, 0.7, 0.168, 0.56, and 4.7 are explained in Figure 8.9 and Table 8.3. From the figure, the PDF file encryption takes more time, i.e., 4.7 milliseconds, while other files consumed only below 1 milliseconds time duration.

This research work is designed in such a way that the user can choose a varied task to encrypt text file, or images or the multimedia files through a single application model. Another important feature of the proposed research work is that varied size of a key for the encryption also allows the users to choose so that the chance of attacks is less since the intruder is unaware of which pair of keys is used for the encryption. This makes the application model more suitable for the encryption process when outsourced in the Cloud. Cipher block chaining is employed and enhanced in this account, presenting an improved method based on the conventional mode of encryption operations. This method is further augmented by integrating the Particle Swarm Optimization (PSO) algorithm for key generation. One interesting feature of

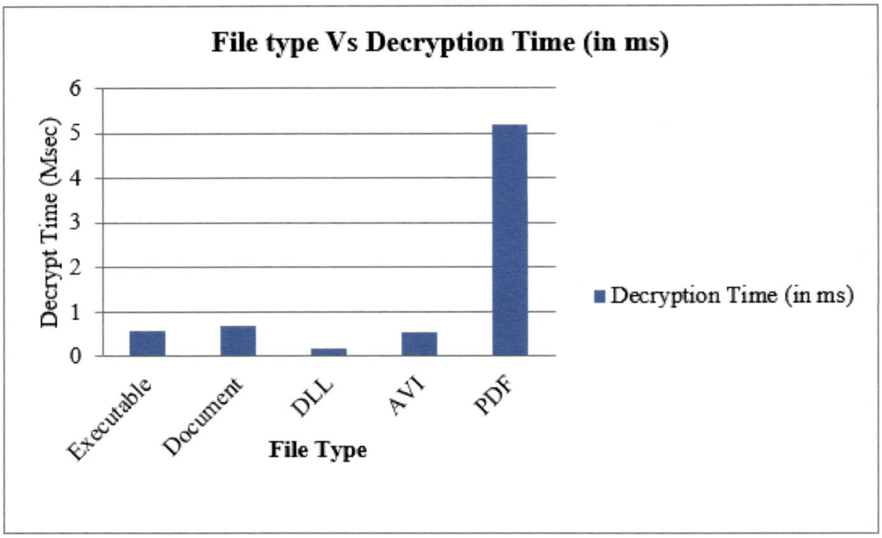

Figure 8.9 Comparisons for various file types vs. decryption time (in M.secs) in NECPSO.

this proposed algorithm is that NECPSO generates a sub-optimal key that makes the process computationally hard. To justify the term sub-optimal, the key generated from the PSO algorithm produces a nearby optimized key for the encryption and thus improves the strength of the algorithm.

8.5 Conclusion

This proposed research discusses the secure communication system in IoT, which is the next revolution in networking and data analysis. To offer a useful security feature, various security-based approaches evolved currently, which are mainly classified into cryptographic based and non-cryptographic based. Both of them have their advantages and limitations; however, adopting a cryptographic-based approach is frequently used. From this viewpoint, the proposed study formulates a strategy to identify and mitigate the unsolved problems associated with IoT's security aspect. A brief discussion of this is carried out in Chapter 1, where it is briefed that the proposed study introduces an intelligent framework for offering secure communication in an IoT. With the aid of distinct research objectives, the proposed system mainly focuses on applying blockchain technology for improving the security aspect of an IoT. The chapter also briefs about the motivational factor for prompting to carry out this topic for investigation. This research illustrates stepwise operations involved in the three distinct algorithm designs, viz. (i) algorithm for reading the flow stream in IoT, (ii) algorithm for the identification of weak port, and (iii) algorithm for the detection and prevention of intrusion. Finally, the implementation carried out toward the algorithm implementation in the form of simulation study has been discussed in this research work. Discussion of the strategy used: outcome obtained concerning performance parameters are carried out and study outcome is benchmarked and found to offer a significant resistance from any form of an unknown attacker without affecting its data forwarding performance in an IoT. The proposed study can be further investigated concerning a certain application that is both time- and mission-critical to justify the applicability of the proposed system on practical grounds. Apart from this, the legitimacy of the software agent is another challenge in software-defined networking based IoT networks. This issue can be addressed by developing a simplified architecture considering critical applications' supportability.

References

[1] Abomhara, M., & Køien, G. M. (2015). Cyber security and the internet of things: vulnerabilities, threats, intruders and attacks. *Journal of Cyber Security and Mobility*, 65-88.

[2] Adil, M., Jan, M. A., Mastorakis, S., Song, H., Jadoon, M. M., Abbas, S., & Farouk, A. (2021). Hash-mac-dsdv: Mutual authentication for intelligent iot-based cyber-physical systems. *IEEE Internet of Things Journal.*

[3] Aleshi, A., Seker, R., & Babiceanu, R. F. (2019, November). Blockchain model for enhancing aircraft maintenance records security. In *2019 IEEE International Symposium on Technologies for Homeland Security (HST)* (pp. 1-7). IEEE.

[4] Awan, I. A., Shiraz, M., Hashmi, M. U., Shaheen, Q., Akhtar, R., & Ditta, A. (2020). Secure framework enhancing AES algorithm in cloud computing. *Security and Communication Networks, 2020.*

[5] Bandyopadhyay, D., & Sen, J. (2011). Internet of things: Applications and challenges in technology and standardization. *Wireless personal communications, 58*(1), 49-69.

[6] Biswas, S., Sharif, K., Li, F., Nour, B., & Wang, Y. (2018). A scalable blockchain framework for secure transactions in IoT. *IEEE Internet of Things Journal, 6*(3), 4650-4659.

[7] Chen, C., Zhang, Z., & Feng, D. (2011, October). Efficient ciphertext policy attribute-based encryption with constant-size ciphertext and constant computation-cost. In *International Conference on Provable Security* (pp. 84-101). Springer, Berlin, Heidelberg.

[8] Duy, P. T., Do Hoang, H., Khanh, N. B., & Pham, V. H. (2019, December). Sdnlog-foren: Ensuring the integrity and tamper resistance of log files for sdn forensics using blockchain. In *2019 6th NAFOSTED Conference on Information and Computer Science (NICS)* (pp. 416-421). IEEE.

[9] Hu, B., Guan, Z. H., Xiong, N., & Chao, H. C. (2018). Intelligent impulsive synchronization of nonlinear interconnected neural networks for image protection. *IEEE Transactions on Industrial Informatics*, *14*(8), 3775-3787.

[10] Kalimuthu, S., Naït-Abdesselam, F., & Jaishankar, B. (2021). Multimedia Data Protection Using Hybridized Crystal Payload Algorithm With Chicken Swarm Optimization. In *Multidisciplinary Approach to Modern Digital Steganography* (pp. 235-257). IGI Global.

[11] Kalimuthu, S. (2021). Sentiment Analysis on Social Media for Emotional Prediction during COVID-19 Pandemic Using Efficient Machine Learning Approach. *Computational Intelligence and Healthcare Informatics*, 215.

[12] Kettani, H., & Wainwright, P. (2019, March). On the top threats to cyber systems. In *2019 IEEE 2nd international conference on information and computer technologies (ICICT)* (pp. 175-179). IEEE.

[13] Lin, J., Yu, W., Zhang, N., Yang, X., Zhang, H., & Zhao, W. (2017). A survey on internet of things: Architecture, enabling technologies, security and privacy, and applications. *IEEE internet of things journal*, *4*(5), 1125-1142.

[14] Mahmud, R., Kotagiri, R., & Buyya, R. (2018). Fog computing: A taxonomy, survey and future directions. In *Internet of everything* (pp. 103-130). Springer, Singapore.

[15] Nawir, M., Amir, A., Yaakob, N., & Lynn, O. B. (2016, August). Internet of Things (IoT): Taxonomy of security attacks. In *2016 3rd International Conference on Electronic Design (ICED)* (pp. 321-326). IEEE.

[16] Rajesh, S., Paul, V., Menon, V. G., & Khosravi, M. R. (2019). A secure and efficient lightweight symmetric encryption scheme for transfer of text files between embedded IoT devices. *Symmetry*, *11*(2), 293.

[17] Rathee, G., Balasaraswathi, M., Chandran, K. P., Gupta, S. D., & Boopathi, C. S. (2021). A secure IoT sensors communication in industry 4.0 using blockchain technology. *Journal of Ambient Intelligence and Humanized Computing*, *12*(1), 533-545.

[18] Roman, R., Zhou, J., & Lopez, J. (2013). On the features and challenges of security and privacy in distributed internet of things. *Computer Networks*, *57*(10), 2266-2279.

[19] Rizvi, S., Kurtz, A., Pfeffer, J., & Rizvi, M. (2018, August). Securing the Internet of Things (IoT): A security taxonomy for IoT. In *2018 17th IEEE International Conference On Trust, Security And Privacy In Computing And Communications/12th IEEE International Conference On Big Data Science And Engineering (TrustCom/BigDataSE)* (pp. 163-168). IEEE.

[20] Shi, P., Song, Q., Patwardhan, J., Zhang, Z., Wang, J., & Gangopadhyay, A. (2019, September). A hybrid algorithm for mineral dust detection using satellite data. In *2019 15th International Conference on eScience (eScience)* (pp. 39-46). IEEE.

[21] Sivanantham, K., Kalaiarasi, I., & Leena, B. (2022). Brain Tumor Classification Using Hybrid Artificial Neural Network with Chicken Swarm Optimization Algorithm in Digital Image Processing Application. *Advance Concepts of Image Processing and Pattern Recognition: Effective Solution for Global Challenges*, 91.

[22] Sivanantham, K. (2022). Deep Learning-Based Convolutional Neural Network with Cuckoo Search Optimization for MRI Brain Tumour Segmentation. In *Computational Intelligence Techniques for Green Smart Cities* (pp. 149-168). Springer, Cham.

[23] Tehrani, M. N., Uysal, M., & Yanikomeroglu, H. (2014). Device-to-device communication in 5G cellular networks: challenges, solutions, and future directions. *IEEE Communications Magazine*, *52*(5), 86-92.

[24] Vimalachandran, P., Liu, H., Lin, Y., Ji, K., Wang, H., & Zhang, Y. (2020). Improving accessibility of the Australian My Health Records while preserving privacy and security of the system. *Health Information Science and Systems*, *8*(1), 1-9.

[25] Viriyasitavat, W., Da Xu, L., Bi, Z., & Hoonsopon, D. (2019). Blockchain technology for applications in internet of things—mapping from system design perspective. *IEEE Internet of Things Journal*, *6*(5), 8155-8168.

[26] Wang, J., Liu, Y., Niu, S., Song, H., Jing, W., & Yuan, J. (2021). Blockchain enabled verification for cellular-connected unmanned aircraft system networking. *Future Generation Computer Systems*, *123*, 233-244.

[27] Wei, J., Liu, W., & Hu, X. (2016). Secure data sharing in cloud computing using revocable-storage identity-based encryption. *IEEE Transactions on Cloud Computing*, *6*(4), 1136-1148.

[28] Wolfert, S., van Wassenaer, L., van der Burg, S., Ryan, M., Klerkx, L., Rijswijk, K., ... & Beers, G. (2021). *Navigating the Twilight Zone: Pathways towards digital transformation of food systems*. Wageningen University & Research.

[29] Xie, L., Ding, Y., Yang, H., & Wang, X. (2019). Blockchain-based secure and trustworthy Internet of Things in SDN-enabled 5G-VANETs. *IEEE Access*, *7*, 56656-56666.

[30] Yang, Y., Guo, Z., Chen, W., Wang, X., Cao, M., Han, X., ... & Qiu, Z. (2021). M2 macrophage-derived exosomes promote angiogenesis and growth of pancreatic ductal adenocarcinoma by targeting E2F2. *Molecular Therapy*, *29*(3), 1226-1238.

[31] Zheng, B. K., Zhu, L. H., Shen, M., Gao, F., Zhang, C., Li, Y. D., & Yang, J. (2018). Scalable and privacy-preserving data sharing based on blockchain. *Journal of Computer Science and Technology*, *33*(3), 557-567.

[32] Zhang, Z. K., Cho, M. C. Y., Wang, C. W., Hsu, C. W., Chen, C. K., & Shieh, S. (2014, November). IoT security: ongoing challenges and research opportunities. In *2014 IEEE 7th international conference on service-oriented computing and applications* (pp. 230-234). IEEE.

9

BlockXAI: Challenges and Future

Akansha Singh[1], Krishna Kant Singh[2], and Mohamed Abouhawwash[3]

[1]SCSET, Bennett University, India
[2]Delhi Technical Campus, India
[3]Department of Mathematics, Faculty of Science, Mansoura University, Mansoura, Egypt
Email: akanshasing@gmail.com

Abstract

Blockchain and explainable AI (XAI) are two rapidly evolving technologies that have the potential to bring significant benefits in various industries and applications. The integration of blockchain and XAI, or BlockXAI, can provide a secure and transparent platform for AI systems and promote trust and adoption of these technologies. However, the integration of blockchain and XAI also faces several challenges, including scalability, interoperability, explainability, privacy, regulation, adoption, and decentralization. To realize the full potential of BlockXAI, it is important to address these challenges and conduct further research to develop new and innovative solutions. The future of BlockXAI is promising, and it has the potential to shape the future of AI by increasing trust, transparency, security, and adoption. However, it is crucial to address the technical, regulatory, and societal challenges that come with these technologies to realize their full potential.

9.1 Introduction

Blockchain is a distributed ledger technology that enables secure and transparent record-keeping of transactions. It operates on a decentralized network of nodes where each node has a copy of the entire ledger and transactions are verified through consensus. This makes blockchain immutable, resistant to tampering and hacking, and enables trust without the need for intermediaries [1].

Explainable AI (XAI) is a field of AI research and development focused on making artificial intelligence systems transparent and interpretable. The goal of XAI is to provide insights into how AI systems make decisions, so that they can be trusted and adopted more widely. This is particularly important in areas such as finance, healthcare, and criminal justice where the impact of AI decisions can be significant. XAI aims to ensure that AI systems are understandable and trustworthy, so that their decisions can be audited and challenged if necessary.

The integration of blockchain and explainable AI (XAI) has the potential to enhance transparency, accountability, and trust in AI systems. By combining blockchain's ability to create a tamper-proof and transparent ledger of transactions with XAI's focus on interpretability and understanding, we can create AI systems that are more trustworthy and reliable.

For example, in a blockchain-based XAI system, the entire decision-making process, data inputs, and outcomes can be recorded on the blockchain. This makes it possible to audit and verify the workings of the AI system, ensuring that it operates within ethical and legal guidelines. Additionally, this type of system could also allow stakeholders to challenge decisions made by AI algorithms and have the possibility for human intervention if necessary.

In conclusion, the integration of blockchain and XAI can bring greater transparency and accountability to AI systems, which can foster trust and increase their adoption across various industries.

9.2 Challenges in Blockchain

There are several challenges that hinder the widespread adoption and implementation of blockchain technology:

- **Scalability:** As more users join a blockchain network, its transaction processing capacity becomes a bottleneck. This can lead to slower transaction times and higher fees.
- **Interoperability:** Different blockchain networks use different protocols and standards, making it difficult for them to communicate and transact with each other.
- **Regulation:** The legal and regulatory landscape for blockchain is still evolving, and there is a lack of clarity on how different jurisdictions will treat it.

- **Security:** Blockchains are decentralized and secure, but they are still vulnerable to attacks such as 51% attacks, smart contract exploits, and phishing scams.
- **Adoption:** Despite its potential, blockchain technology has not yet been widely adopted due to its complexity, lack of understanding, and the need for significant changes in existing systems and processes.
- **Energy consumption:** Blockchain consensus algorithms such as Proof-of-Work require a significant amount of energy and computational resources, which can make them unsustainable in the long term.
- **Limited use cases:** While blockchain has a wide range of potential use cases, it is still in its early stages and has not yet been proven to be the best solution for many of the problems it is intended to solve.

Overall, the challenges facing blockchain technology can be overcome through research, innovation, and collaboration. However, it will take time for the technology to mature and for the challenges to be resolved.

9.3 Challenges of XAI

There are several challenges associated with explainable AI (XAI):

- **Complexity:** AI systems can be highly complex and their internal workings can be difficult to understand. This makes it challenging to explain the decision-making process of AI algorithms and the reasoning behind their output.
- **Data bias:** AI models are only as good as the data they are trained on. If the data used to train the model is biased, the AI system will also be biased, and its outputs will not be reliable.
- **Limitations of interpretability methods:** There are various interpretability methods available to explain the workings of AI systems, but they are often limited in scope and may not provide a complete understanding of the system's decision-making process.
- **Lack of transparency:** Many AI systems are proprietary and their workings are not disclosed to the public, making it difficult to understand their decision-making processes and to assess their fairness and accuracy.
- **Trade-off between explainability and accuracy:** There is often a trade-off between the interpretability of an AI system and its accuracy. The more complex and accurate the AI system, the less interpretable it may be.

- **Validation and verification:** The explanations provided by XAI methods need to be validated and verified to ensure that they are accurate and that the AI system is operating as expected.
- **Ethical considerations:** AI systems can have a significant impact on people's lives, and it is important to ensure that they are fair, ethical, and transparent. XAI can play a role in promoting accountability and transparency in AI systems, but there are also ethical considerations that need to be taken into account.

Despite these challenges, XAI is a rapidly growing field and there is significant research and development taking place to address these challenges and make AI systems more explainable and trustworthy.

9.4 Future of Blockchain

The future of blockchain technology is uncertain but holds a lot of potential. Here are some of the ways in which blockchain is expected to evolve in the coming years:

- **Widespread adoption:** Blockchain is expected to become more widely adopted across a range of industries, including finance, supply chain management, healthcare, and government [2].
- **Increased scalability:** Efforts are underway to increase the scalability of blockchain technology to handle more transactions and users, making it more feasible for widespread adoption.
- **Interoperability:** Efforts are also underway to make different blockchain networks more interoperable, allowing them to communicate and transact with each other.
- **Improved security:** The security of blockchain technology is expected to improve as researchers identify and address vulnerabilities and new security measures are developed.
- **Regulatory clarity:** As the legal and regulatory landscape for blockchain continues to evolve, it is expected that there will be greater clarity on how it will be treated by different jurisdictions.
- **Decentralized finance:** Decentralized finance (DeFi) is expected to grow in popularity, with more and more people using blockchain-based financial instruments such as stablecoins and decentralized exchanges.
- **New use cases:** As blockchain technology continues to mature, new and innovative use cases are expected to emerge, leveraging its decentralized and secure nature.

Overall, the future of blockchain is exciting and holds a lot of potential, but it will also be shaped by the challenges it faces and the progress made in addressing these challenges.

9.5 Future of XAI

The future of explainable AI (XAI) is rapidly evolving and holds a lot of potential. Here are some of the ways in which XAI is expected to evolve in the coming years:

- **Widespread adoption:** XAI is expected to become more widely adopted across a range of industries, including finance, healthcare, and government, where transparency and accountability in AI decision-making are critical.
- **Improved interpretability methods:** Efforts are underway to improve the interpretability of AI systems, making it easier to understand their decision-making processes and the reasoning behind their outputs [3].
- **Integration with other technologies:** XAI is expected to be integrated with other technologies such as blockchain, enabling secure and transparent decision-making processes.
- **Increased trust in AI:** As XAI becomes more widespread and its capabilities continue to improve, it is expected to increase trust in AI systems and their decision-making processes.
- **Ethical considerations:** As XAI becomes more prevalent, there will be a need to address ethical considerations, such as ensuring that AI systems are fair and unbiased.
- **New use cases:** As XAI technology continues to mature, new and innovative use cases are expected to emerge, leveraging its ability to provide transparency and accountability in AI decision-making.

Overall, the future of XAI is promising and holds a lot of potential, but it will also be shaped by the challenges it faces and the progress made in addressing these challenges.

9.6 Integration of BlockXAI

Blockchain technology can help explainable AI by providing a secure and transparent platform for storing and accessing the data used by AI systems [3]. Here are a few ways in which blockchain can help improve the explainability of AI:

- **Tamper-resistant data storage:** Blockchain's decentralized and cryptographic nature makes it difficult for data to be tampered with, ensuring that the data used by AI systems is trustworthy.
- **Traceability of decisions:** By storing information on the blockchain, it becomes possible to trace the data used to make decisions and the steps taken by the AI system to arrive at a particular outcome.
- **Increased transparency:** Blockchain technology provides a transparent platform for storing and accessing data, making it easier to understand how AI systems make decisions.
- **Audibility:** Blockchain technology allows for audibility, as all transactions are stored in a tamper-proof ledger that can be audited to ensure that AI systems are making decisions in a fair and transparent manner.
- **Improved accountability:** By making it possible to track the data and decision-making processes of AI systems, blockchain technology can improve accountability and increase trust in AI systems.

Overall, the integration of blockchain technology with explainable AI has the potential to provide a secure and transparent platform for AI decision-making, allowing for greater accountability and trust in these systems.

Explainable AI (XAI) can help improve the functioning and trustworthiness of blockchain systems in several ways:

1. **Improved decision-making:** XAI can be used to improve the decision-making processes of smart contracts, ensuring that they are transparent and free from biases.
2. **Better understanding of blockchain systems:** By providing explanations for the behavior of blockchain systems, XAI can help improve the overall understanding and trust in these systems [3].
3. **Enhanced security:** XAI can be used to identify and address security vulnerabilities in blockchain systems, helping to prevent hacking and other forms of malicious activity.
4. **Increased transparency:** By providing explanations for the behavior of blockchain systems, XAI can increase transparency and help build trust in these systems.
5. **Improved scalability:** XAI can help improve the scalability of blockchain systems by identifying inefficiencies and bottlenecks, allowing for better optimization and improved performance.

Overall, the integration of explainable AI with blockchain technology can help improve the functioning and trustworthiness of blockchain systems, making them more secure, transparent, and scalable.

9.7 Challenges in Integrating Blockchain with XAI

The integration of blockchain and explainable AI (XAI) faces several challenges, including:

1. **Technical complexity:** Integrating blockchain and XAI requires advanced technical knowledge and expertise, as both technologies are complex in their own right.
2. **Interoperability:** Ensuring interoperability between different blockchain platforms and AI systems can be a challenge, as they use different protocols and standards [4].
3. **Data privacy:** Storing sensitive and personal data on a blockchain requires robust privacy protection measures, which can be challenging to implement in practice.
4. **Scalability:** Both blockchain and XAI systems can experience scalability issues, particularly in the context of large-scale applications.

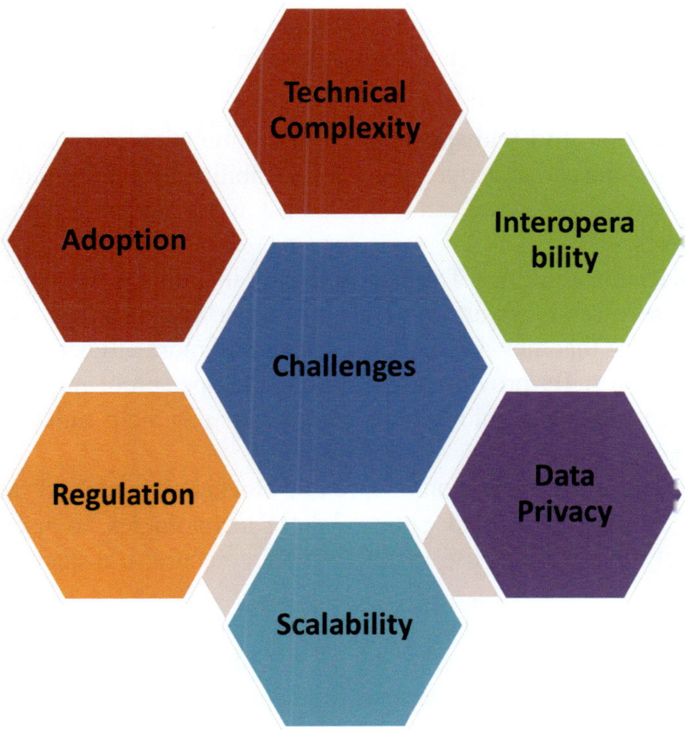

5. **Regulation:** The integration of blockchain and XAI is still a relatively new field and there are concerns about how these systems will be regulated and governed.
6. **Adoption:** Despite the potential benefits, the integration of blockchain and XAI faces challenges in terms of adoption, as users may be hesitant to embrace new and unfamiliar technologies.
7. **Explainability:** XAI systems can be complex and difficult to understand, making it challenging to explain their decision-making processes to users.

Despite these challenges, the integration of blockchain and XAI has the potential to bring significant benefits in terms of security, transparency, and accountability, and many experts believe that it will play a key role in shaping the future of AI.

9.8 Road Ahead: BlockXAI

The future of blockchain with explainable AI (XAI) is promising and has the potential to bring significant benefits in various industries and applications [5]. Here are a few ways in which the integration of blockchain and XAI can shape the future:

- **Improved trust and transparency:** The combination of blockchain's secure and transparent ledger with XAI's ability to explain complex decision-making processes can increase trust in AI systems and promote transparency.
- **Better decision-making:** The use of XAI can improve the decision-making processes of smart contracts and other decentralized applications, ensuring that they are free from biases and more efficient.
- **Better security:** By providing a secure and tamper-proof platform for storing sensitive data, blockchain technology can help enhance the security of AI systems.
- **Increased adoption:** The combination of blockchain and XAI has the potential to increase the adoption of these technologies, as users are more likely to trust and use systems that are secure, transparent, and explainable.
- **Improved compliance:** The transparency and traceability provided by blockchain, combined with the explainability of XAI, can help organizations comply with various regulations, such as data privacy laws.

Overall, the integration of blockchain and XAI has the potential to shape the future of AI by increasing trust, transparency, security, and adoption [6]. However, it is important to address the technical, regulatory, and societal challenges that come with these technologies to realize their full potential.

9.9 Legal and Ethical Issues in BlockXAI

The integration of blockchain and explainable AI (XAI) brings new legal and ethical challenges and issues that need to be considered and addressed:

- **Privacy:** The storage of sensitive and personal data on a blockchain and the use of XAI algorithms for decision-making processes raise privacy concerns, and there is a need for clear privacy laws and regulations [7].
- **Data protection:** With the increasing use of XAI algorithms, there is a need for clear and comprehensive data protection laws that ensure that personal data is not used for discriminatory purposes.
- **Bias and discrimination:** The use of AI algorithms in decision-making processes can lead to biased and discriminatory outcomes, and there is a need for regulations and tools to mitigate these risks [8].
- **Responsibility:** The decentralization and automation of processes in blockchain systems raises questions about who is responsible in the event of errors or unintended consequences.
- **Intellectual property:** Intellectual property laws and regulations will need to be adapted to accommodate the unique features of blockchain and XAI.
- **Decentralization:** Decentralization is a key aspect of blockchain, and there are legal and ethical challenges that come with the decentralization of decision-making processes [9].
- **Regulation:** The integration of blockchain and XAI is still a relatively new field, and there is a need for clear and comprehensive regulations that ensure that these technologies are used in a responsible and ethical manner.

These legal and ethical issues represent important challenges that need to be addressed to ensure that the integration of blockchain and XAI is beneficial for society as a whole. It is important for governments, organizations, and individuals to work together to develop clear laws, regulations, and ethical standards for these technologies [10].

9.10 Open Research Problems in BlockXAI

There are several open research problems in the field of blockchain and explainable AI (BlockXAI):

1. **Scalability:** Scalability is a major challenge in both blockchain and XAI, and finding ways to make these systems more scalable and efficient remains an important open research problem.
2. **Interoperability:** Ensuring interoperability between different blockchain platforms and AI systems is a significant challenge, and more research is needed to develop standards and protocols for interoperability.
3. **Explainability:** XAI systems can be complex and difficult to understand, and more research is needed to develop explainable AI systems that are more transparent and accessible to users [11].
4. **Privacy:** Storing sensitive and personal data on a blockchain requires robust privacy protection measures, and more research is needed to develop privacy-preserving methods for BlockXAI.
5. **Regulation:** The integration of blockchain and XAI is still a relatively new field, and more research is needed to understand how these systems will be regulated and governed in the future.
6. **Adoption:** Despite the potential benefits, the integration of blockchain and XAI faces challenges in terms of adoption, and more research is needed to understand how to increase user acceptance and trust in these systems.
7. **Decentralization:** Decentralization is a key aspect of blockchain, and more research is needed to understand how to balance the trade-off between decentralization and efficiency in BlockXAI systems.

These open research problems represent exciting opportunities for researchers and practitioners to advance the field of BlockXAI and develop new and innovative solutions for these challenges.

9.11 Conclusion

In conclusion, the integration of blockchain and explainable AI (XAI), or BlockXAI, has the potential to shape the future of AI by providing a secure and transparent platform for AI systems and promoting trust and adoption of these technologies. BlockXAI has the potential to solve some of the biggest challenges facing AI, such as privacy, data protection, bias and discrimination, and responsibility. However, it is important to address the technical, legal, and ethical challenges that come with these technologies to ensure that they are used in a responsible and ethical manner. The future of BlockXAI is promising, but it is crucial to conduct further research and development to address the open research problems and legal and ethical issues facing these technologies. By working together to address these challenges, the integration of blockchain and XAI has the potential to bring significant benefits to society as a whole.

References

[1] S. Nakamoto, *Bitcoin: A peer-to-peer electronic cash system*, 2008, [online] Available: https://bitcoin.org/bitcoin.pdf.

[2] D. Tapscott and A. Tapscott, *Blockchain revolution*, 2016, [online] Available: https://blockchain-revolution.com/.

[3] Narayanan, J. Bonneau, E. Felten, A. Miller and S. Goldfeder, Bitcoin and Cryptocurrency Technologies: A Comprehensive Introduction, Princeton, NJ, USA:Princeton Univ. Press, 2016.

[4] Singh, A., & Singh, K. K. (2023). Blockchain with Federated Learning for Secure Healthcare Applications. *Blockchain and Deep Learning for Smart Healthcare*, 35-44.

[5] M. Antonopoulos and G. Wood, Mastering Ethereum: Building Smart Contracts and Dapps, Sebastopol, CA, USA:O'Reilly Media, 2018.

[6] L. Lantz and D. Cawrey, Mastering Blockchain, Sebastopol, CA, USA:O'Reilly Media, 2020.

[7] Bashir, Mastering Blockchain, Birmingham, U.K.:Packt Publishing, 2020.

[8] Singh, A., & Singh, K. K. (2023). Blockchain and Deep Learning: Research Challenges, Open Problems, and Future. Blockchain and Deep Learning for Smart Healthcare, 425-431.

[9] Rakhra, A., Gupta, R., & Singh, A. (2021). Blockchain and Internet of Things Across Industries. Machine Learning Approaches for Convergence of IoT and Blockchain, 1-34.

[10] M. M. Ghonge, N. Pradeep, R. V. Ravi and R. Mangrulkar, "A comprehensive review of the security and privacy issues in blockchain technologies" in Blockchain Technologies and Applications for Digital Governance, Hershey, PA, USA:IGI Global, pp. 180-199, 2022.

[11] Singh, A., & Singh, K. K. (2024). Blockchain and Deep Learning for Smart Healthcare. John Wiley & Sons.

Index

About the Editors

Prof. Akansha Singh, Professor at the School of Computer Science and Engineering, Bennett University, Greater Noida, boasts a comprehensive academic background with a B.Tech, M.Tech, and Ph.D. in Computer Science. Her doctoral studies, conducted at the prestigious IIT Roorkee, were focused on the cutting-edge fields of image processing and machine learning. A prolific author and scholar, Dr. Singh has contributed over 100 research papers and penned several books, alongside a multitude of conference papers. Her editorial expertise is recognized by leading publishers such as Elsevier, Taylor and Francis, and Wiley, where she has edited books on a variety of emerging topics. Her research interests are diverse and influential, spanning image processing, remote sensing, the Internet of Things (IoT), Blockchain and machine learning. Prof. Singh's work in these areas not only advances the field of computer science but also significantly contributes to the broader scientific and technological community.

Prof. Dr. Krishna Kant Singh, currently the esteemed Director of Delhi Technical Campus in Greater Noida, India, is a highly experienced educator and researcher in the field of engineering and technology. His academic journey encompasses a B.Tech, M.Tech, and a Ph.D. from the renowned Indian Institute of Technology (IIT) Roorkee, specializing in image processing and machine learning. Dr. Singh's scholarly contributions are remarkable, with over 150 research papers published in prestigious Scopus and SCIE indexed journals. His commitment to knowledge dissemination is further evidenced by his authorship of 25 technical books, which serve as significant resources in the field of engineering and technology. Dr. Singh's dedication to his field is evident through his involvement in various capacities, from research and writing to editorial responsibilities. His work not only advances the field of engineering and technology but also inspires future generations of engineers and researchers.

For Product Safety Concerns and Information please contact our
EU representative GPSR@taylorandfrancis.com Taylor & Francis
Verlag GmbH, Kaufingerstraße 24, 80331 München, Germany